PRAIRIE STATE BOOKS

In conjunction with the Illinois Center for the Book, the University of Illinois Press is reissuing in paperback works of fiction and nonfiction that are, by virtue of authorship and/or subject matter, of particular interest to the general reader in the state of Illinois.

Publication of Prairie State Books is supported by a generous grant from William and Irene Beck.

A list of books in the series appears at the end of this book.

THE DRUMS OF THE 47th

Robert J. Burdette

THE DRUMS OF THE 47th

Robert J. Burdette
Introduction by John E. Hallwas

University of Illinois Press
Urbana and Chicago

Manufactured in the United States of America
♾ This book is printed on acid-free paper.

Frontispiece photograph from *Robert J. Burdette, His Message,*
ed. Clara B. Burdette (Clara Vista Press, 1922)

Library of Congress Cataloging-in-Publication Data
Burdette, Robert J. (Robert Jones), 1844–1914.
The drums of the 47th / Robert J. Burdette ;
introduction by John E. Hallwas.
p. cm.—(Prairie State books)
Originally published : Indianapolis: Bobbs-Merrill, c1914.
With a new introduction.
Includes bibliographical references.
ISBN 978-0-252-06853-9 (acid-free paper)
1. Burdette, Robert J. (Robert Jones), 1844–1914.
2. United States. Army. Illinois Infantry Regiment, 47th (1861–1864)
3. United States—History—Civil War, 1861–1865 Personal narratives.
4. Illinois—History—Civil War, 1861–1865 Personal narratives.
5. United States—History—Civil War, 1861–1865—Regimental histories.
6. Illinois—History—Civil War, 1861–1865—Regimental histories.
7. Soldiers—Illinois—Peoria Biography.
I. Title. II. Series.
E505.5 47th.B87 2000
973.7'473—dc21 99-36036
CIP

P 6 5 4 3 2

CONTENTS

INTRODUCTION

John E. Hallwas

Robert J. Burdette's Civil War memoir, *The Drums of the 47th,* has been largely ignored by historians. Bell Irvin Wiley, for example, in his classic study, *The Life of Billy Yank: The Common Soldier of the Union* (1952), never refers to it in his text and does not list it in his bibliography. Neither does Victor Hicken, in his highly praised book, *Illinois in the Civil War* (1966; 2d ed., 1991), although he lists more than ten other sources for the 47th Infantry Regiment. There are perhaps good reasons for omitting it: *The Drums of the 47th* offers little information about specific people, places, and events, and it does not provide a chronological account of Burdette's experience. Strictly speaking, the book has little historical value.

Nevertheless, it is a distinctive and even remarkable work that deserves a modern readership.[1] Few Civil War accounts have the literary quality of Burdette's memoir, which effectively depicts the inner life of the common soldier as it portrays selected aspects of the author's experience in the great conflict. *The Drums of the 47th* is, in fact, one of the few Civil War accounts by a professional writer, for Burdette was a well-known humorist, the author of poems and prose sketches that were widely admired in his own time.

Born in Greensboro, Pennsylvania, on July 30, 1844, he moved with his family to Cummingsville, Ohio, near Cincinnati, two years later. He was one of ten children, two of whom died as infants. In

1852 the Burdettes moved again, to Peoria, which was then a rapidly growing river town of six or seven thousand people. There Robert attended the Hinman school, which he later depicted in a widely reprinted newspaper account of schoolboy rebellion, "The Strike at Hinman's." He also attended Peoria High School, graduating in December 1861.

Young Burdette wanted to join the Union army, but after much talking with his parents he agreed to wait until his eighteenth birthday. On August 4, 1862, just five days after he turned eighteen, he enlisted in Company C of the 47th Infantry Regiment. It was one of four units that made up the "Eagle Brigade," so named because one of them, the 8th Wisconsin Infantry Regiment, brought along a bald eagle mascot known as "Old Abe," which soon became famous. Burdette's regiment fought in many small battles and skirmishes and participated in the Siege of Vicksburg. During his three years of service, the cheerful, bookish, patriotic youth remained a private but was distinguished for bravery, and he was discharged at Selma, Alabama, on July 20, 1865, shortly before his twenty-first birthday.

After the war, he worked at several jobs, including teaching, in and around Peoria. Following his mother's death in 1868, he left to study art and languages at the famous Cooper Institute in New York. His frequent letters to the *Peoria Transcript* led to the launching of his literary career. In the fall of 1869 the *Transcript*'s editor, Enoch Emery, asked Burdette to return to Peoria and write for the paper. He did so, but Emory's lack of interest in his humorous sketches prompted him to switch, in 1870, to the *Peoria Review,* where his career as a humorist began.

When that newspaper failed in 1874, he was hired by the *Burlington Hawkeye,* where he was the city editor and, later, the managing editor. Burdette started a biweekly feature titled "Hawkeyetems of Roaming Robert, the Hawkeye Man," and those humorous columns, often set in Burlington and focused on small-town character types, were soon widely reprinted, making the *Hawkeye* known across the country. In 1876 Burdette also became a platform speaker, and his first lecture, "The Rise and Fall of the Mustache"—a humorous commentary on the transition from youth to manhood—soon became very popular. In fact, it became one of the most widely acclaimed lectures of the era. It led to his first book, *The Rise and Fall of the Mustache and Other "Hawkeyetems"* (1877), which includes his long lecture and collects dozens of short prose and poetry items that he had written for the *Hawkeye.*

Burdette continued as a successful writer and lecturer for many years. He published three subsequent prose collections, *Hawk-eyes* (1879)—later reissued as *Innach Garden and Other Comic Sketches* (1886) and *Schooners That Pass in the Dark* (1894)—*Chimes from a Jester's Bells* (1897), and *Old Tom and Young Tom* (1912), as well as two volumes of poetry, *Smiles Yoked with Sighs* (1900) and *The Silver Trumpets* (1912). He also wrote a biography of William Penn, published in 1892, and a life of Baptist minister Russell H. Conwell, published in 1894. Burdette's platform career brought him into contact with other noted humorists, including Eugene Field, Bill Nye, James Whitcomb Riley, Henry W. Shaw ("Josh Billings"), and Mark Twain. Riley became a close friend.

Burdette left Burlington in 1880 and eventually settled in Ardmore, Pennsylvania, but he continued to write for the *Hawkeye* as a correspondent. He also wrote a regular column for the *Brooklyn Eagle* and occasional items for the *Philadelphia Press.* Before long his newspaper pieces were syndicated across the country, and he had a considerable national reputation. After his invalid wife, Carrie, died in 1884, he started working as a pastor at the Lower

Merion Baptist church in Bryn Mawr. Although he continued to travel as a lecturer, Burdette soon preached widely as well, becoming a licensed minister in 1887.

His later years were spent in California. In 1899 he accepted a call to preach at the First Presbyterian Church of Pasadena—although he remained a Baptist—and he also married his second wife, a widow named Clara Baker. Four years later he became the pastor of the Temple Baptist Church in Los Angeles, where he was ordained in 1903. That church grew rapidly under Burdette's leadership, often drawing a Sunday attendance of more than three thousand, and he also wrote for the *Los Angeles Times Magazine* and lectured frequently in California.

In 1909 a fall at his home caused an injury that led to chronic pancreatitis, so Burdette resigned his pastorate, although he continued to preach from time to time at various locations. Early in 1911 he was asked to write a series of Civil War reminiscences for the *Sunday School Times,* and those articles, published from the fall of 1911 to the summer of 1912, were eventually combined with two other short memoirs to produce *The Drums of the 47th.* Burdette's health steadily deteriorated, and a few days after the book appeared he died, on November 19, 1914.

Even before his death, Burdette's literary reputation was declining, and consequently his books have been out of print for several decades. In 1922 his widow, Clara, published a biography, *Robert J. Burdette: His Message,* which is the only substantial work about him.[2] Burdette's humorous writings in prose and verse—chiefly gentle satire, local color, character types, and ridicule of cultural myths—are not apt to interest modern readers, but *The Drums of the 47th* deserves to be widely known.

Those who are interested in historical information about the 47th Infantry Regiment would be better served by reading Byron

INTRODUCTION

Cloyd Bryner's *Bugle Echoes: The Story of the Illinois 47th* (1905), a vigorous and detailed account of that unit, which participated in twenty-six engagements, from March 1862 to December 1864. During that time, it marched some three thousand miles and lost 521 men, including four colonels, from wounds and disease. As Burdette proudly points out in his memoir, it was "a marching and a fighting regiment" (43). Since he maintained contact with some of his comrades and participated in regimental reunions, he surely read *Bugle Echoes,* which mentions him briefly—if only because he became famous after the war.[3]

Because Bryner's book covers the history of the regiment, depicting the officers, travels, and engagements, Burdette must have felt that he did not need to duplicate any of that with his own reminiscences. Also, the *Sunday School Times* apparently influenced the form and focus of *The Drums of the 47th.* Because it was a weekly periodical, Burdette conceived of his autobiographical account as a series of personal essays on various topics related to his war experience, and because it was a religious publication, he portrayed the inner life of the common soldier and occasionally emphasized the Christian perspective.

An article in the October 14, 1911, issue of the *Sunday School Times* announces the start of Burdette's "reminiscences" and prints a letter that he sent along with the first installment. The letter is not important, but with it are lists of possible series titles and proposed chapter titles (or topics), which shed light on the writing of the book. The series titles, printed below, are preceded by a comment to the editors:

> The General Subject: Pick 'em while they last [i.e., choose the title for the series], to suit the editorial council of the *Times.* Any one will suit me, and you should know better than anyone else

> which one will best meet the taste of your—OUR—readers. There are times when I bow to editorial omniscience. . . .
>
> FIFTY YEARS AFTER: A Veteran's Memories of a Recruit's Experience
> THE GAME OF WAR: Reminiscences of a Pawn
> THE GAME OF WAR: Experiences of a Pawn
> WAR AND PEACE: Reviewing the Battle from a Dove-cote
> FIGHTS AND FROLICS: Lights and Shadows of a Soldier's Life
> MARCHING WITH THE MEN: What a Boy Saw in the Civil War[4]

From this list the editors chose *Lights and Shadows of a Soldier's Life,* one of his subtitles, as the title of the serialized work, which began appearing in the October 14 issue. Installments were printed every three or four weeks. The discarded titles shed light on Burdette's approach to his subject matter. For example, "FIFTY YEARS AFTER: A Veteran's Memories of a Recruit's Experience" reveals his awareness that half a century separated himself-as-author from himself-as-soldier, so his account would be marked by a mature, thoughtful perspective. Unlike so many other Civil War memoirs, it would not be simply a report but an inquiry into the meaning of a young soldier's experience. Likewise, his use of the word "Pawn" in two of the subtitles reveals his recognition that he had been simply a minor player—and therefore an expendable figure—in "THE GAME OF WAR." And his final suggestion, "MARCHING WITH THE MEN: What a Boy Saw in the War," emphasizes that his memoir will be an account of maturation, of manhood achieved under highly stressful circumstances.

Burdette's list of proposed "CHAPTER TITLES" (or chapter and section titles, actually) is also interesting but is too long to print here. It is perhaps sufficient to say that some of his topics, such as "Hospital Duty," "The Stately Southern Homes," "The Girl I Left behind Me," and "A Cavalry Charge with Custer," never appeared in *Lights and Shadows of a Soldier's Life,* and others that did appear are not on that list. Also, in some cases his titles were changed, but the topics were covered. His listed chapter title "The Skirmish," for example, became "The Real Thing" in both the series and the book. In any case, Burdette's list of proposed "CHAPTER TITLES" shows that *The Drums of the 47th* was not written in advance of the serialized publication. It took shape as he wrote it over the nine-month period in which *Lights and Shadows of a Soldier's Life* appeared in the *Sunday School Times*—and afterward, when he added two more chapters for the book version. In fact, he tells the editors at the close of his long list of proposed titles, "The order of these will certainly be changed somewhat. And I think I have too many of them. But you can let me know in time to back off the stage gracefully." In short, they can end the serialized account whenever they want, just by letting him know.

The fifteen chapters that comprise *Lights and Shadows of a Soldier's Life* are changed very little—hardly a word or two per chapter—for publication in *The Drums of the 47th.* The one notable exception occurs in the fourteenth chapter, "The Colonel," which in the *Sunday School Times* has an additional, closing paragraph:

> The colonel gives the regiment its character. He influences the regiment as the pastor does his church. Under Bryner, our first and best loved colonel, the regiment was a family. Bryner was more of a father to all the men than any of his successors. Every man loved him and admired him, and he was worthy of it all. Under Cromwell we were boys, even as was that dashing commander; care-free, rollicking, ready at any time for "a fight or a frolic" and not much caring which. Under McClure the

> Forty-seventh reached a climax of discipline and orderly bearing. He made us soldiers. He was a Presbyterian of the steel-bluest, and he "Oliver Cromwellized" the regiment.[5]

Perhaps Burdette later felt that his use of "Oliver Cromwellized" to describe the impact of the colonel whose name was *not* Cromwell created some confusion, but regardless of why he later omitted it, the paragraph is actually a perceptive comment on the psychological impact that colonels often had on their regiments.

The most significant difference between the 1911–12 *Sunday School Times* version and the 1914 book version is that *Lights and Shadows of a Soldier's Life* does not include chapter XV, "A Triptych of the Sixties," and chapter XVII, "The Lost Fort." As a result, the serialized version ends with "The Farewell Volleys," which later becomes the penultimate chapter (XVI) in the book. Because that chapter deals with the funeral and burial of a soldier, it carries a strong Christian message—of anticipated resurrection through Christ—and that was probably why it was chosen to close the periodical version. Since Burdette had informed his editors that they could select what appeared and determine when the series ended, they may have decided that closing with the essential Christian message was the appropriate thing to do in the *Sunday School Times.* Whether they did or not, the serialized version has a more pointed Christian focus because it ends with "The Farewell Volleys."

It is also worth noting that chapter XV in *The Drums of the 47th,* "A Triptych of the Sixties," depicts the execution of three soldiers for raping a mother and daughter. If Burdette had sent that chapter to the editors, they surely would have chosen to omit it and, instead, print his next installment. It is also possible, of course, that Burdette simply withheld "A Triptych of the Sixties" or wrote it later for the book version. Although not objectionable in content, chapter XVII, "The Lost Fort," may also have been omitted by the editors because, as a closing piece, it merely emphasizes the futility of war, whereas "The Farewell Volleys" emphasizes salvation through Christ.

The difference in the endings of the serialized and book versions also points to an important aspect of *The Drums of the 47th*—the tension between Burdette's proud recollection of his military experience and his Christian sensibility, which condemns war. As he says in chapter I, "I went into the army a light-hearted boy" and "I had the rollicking time of my life and came home stronger than an athlete" (9). In contrast, the final chapter makes a powerful antiwar statement, as he asserts that the victory at Spanish Fort, measured in agony, heartache, and blood, simply demonstrated that "We are the best killers" (205). This difference is surely a result of the change in himself, from the period of his youth, when the Civil War was a great adventure through which he achieved manhood, to his later life, when he was a minister interpreting his past from a Christian perspective.

That Burdette was not focused on the facts of his military service but on the meaning of that experience in human terms is apparent at the outset, when he collapses the time between the fall of Fort Sumter in April 1861 and the moment of his enlistment sixteen months later:

> Boom! a siege-gun fired away off down in Charleston. . . . Far away, from the ramparts of Sumter, a bugle shrilled across the states as though it were the voice of the trumpet of the angel calling the sheeted dead to rise. And close at hand the flam, flam, flam of the drum broke into wild thrill of the long roll—the fierce snarl of the dogs of war. . . .

> I leaped to my feet, seized my cap and ran to the window to wind my arms around my mother's neck.
>
> "Mother," I said, "I'm going!" (3–4)

His point, of course, is that he responded to the patriotic atmosphere, symbolized by the drums, that developed after Fort Sumter fell when he decided to enlist. He was, then, typical of thousands of others who responded in exactly the same way during the earlier part of the war. That he had actually waited until the summer of his eighteenth birthday, when Peoria's 47th Regiment was in need of new recruits, is never mentioned, for it is irrelevant to Burdette's portrayal of his response to the war fever that drew him into military service.

Early sections of *The Drums of the 47th* focus on uniforms, music, training, camp stories, and other matters, but Burdette also begins to reflect the inner life of the common soldier, especially in the section titled "The Soldier's Rainy Day Religion." That leads directly to chapter V, "The Murder," which is one of the most striking and effective parts of the book. After introductory remarks that ponder the meaning of "casualty" and suggest that warfare simply legitimizes murder, he presents a vivid account of the first death in battle that he witnessed at close range—"a boy about my own age, not over nineteen" (52):

> The young artilleryman leaped to his feet, his face lifted toward the gray sky, his hands tossed above his head. He reeled, and as a comrade sprang to catch him in his arms the boy cried, his voice shrilling down the line:
>
> "Murder, boys! Murder! Oh, murder!"
>
> He clasped his hands over a splotch of crimson that was widening on the blue breast of his red-trimmed jacket and fell into

> the strong arms of the comrades who carried him to the rear. Him, or—It.
>
> The rain began again and the warm drops fell like tears upon the white faces, as though angels were weeping above him. I watched the men carry him away to where the yellow flag marked the mercy station of the field hospital.
>
> Fear, before unfelt because unknown, clutched my heart like the hand of death.(53)

That is surely one of the most powerful depictions of a casualty in the surviving records of the Civil War, and it is typical of Burdette that he also mentions the impact of that shocking event on him. According to Clara Burdette, the young artilleryman's death occurred at Jackson, Mississippi, in the spring of 1863, but the place and time, never indicated in *The Drums of the 47th,* are irrelevant to the effectiveness of that scene as a literary reconstruction of a terrifying personal experience. In the pages that follow, Burdette depicts his struggle to cope with that awful memory, which haunted his dreams and "marred the glory of victory" (59).

Undoubtedly, one reason for the effectiveness of chapter V is that it was a revision of "Laurel and Cypress," a story that first appeared in *Chimes from a Jester's Bells.* That earlier work is a grimly ironic account in which the author recalls that as a youth he "yearned for battle," for "arms on armor clashing," only to be devastated by witnessing the death of a young artilleryman.[6] Published shortly after the appearance of Stephen Crane's *The Red Badge of Courage,* it may have been influenced by that famous novel, which depicts a similar inner change in the protagonist, Henry Fleming.

In chapters VI–X Burdette ponders the symbolic significance of the flag, the meaning of comradeship, the struggle with fear, and other aspects of the common soldier's experience, and then the book becomes more focused on specific memories. Chapter XI, for example, vividly recounts the fighting at the Battle of Corinth on October 3–4, 1862, and chapter XII describes his brief meeting with General Grant.

Considering that *The Drums of the 47th* does not present a chronological account of Burdette's experience, chapter XVII, "The Lost Fort," closes the book about as well as one section could. It is based on the Battle of Spanish Fort, Alabama, which took place in the early spring of 1865 and brought an end to the combat experience of Burdette and his regiment. The chapter is structured by the development of a single day, moving as it does from "the darkness before the dawn" (197) through the morning hours before a battle, the fighting itself in the afternoon, and the aftermath the following night. More important, it consistently reflects the emotional experience of the soldiers, who are weary, tense, courageous, and determined. The climactic section is a stunning impressionistic account that compresses the fighting into a single paragraph, most of which is a thirty-six-line sentence filled with the sights and sounds of battle. It closes with "fighting men swarming like locusts into the embrasures; saber and bayonet, sponge staff and rammer, lunge, thrust, cut, and crashing blow; men driven out of the embrasures . . . and slashing their way back again like fighting bulldogs, holding every inch they gain; hand to throat and knife to heart; hurrying reinforcements from all sides racing to the crater of smoke and flame; a long wild cheer . . . a white flag fluttering like a frightened dove amidst smoke and flame, the fury and anguish, the hate and terror, the madness and death of the hell of passion raging over the sodden earth—the fort is ours" (204–5). Here, too, *The Drums of the 47th* may show the influence of *The Red Badge of Courage,* which employs literary impressionism to heighten the emotional impact of Henry Fleming's battle experience.

Burdette's vivid description of the fighting—how it seemed to him as a recruit—is followed by a reflection on the impact of the battle, as measured by the suffering of the soldiers and the anguish of their families. The mind of the older Burdette, the veteran, the memoirist, the Christian moralist, is clearly felt as he comments, "The paying for a fort goes on as long as a winner or loser is left alive—heartache and loneliness and longing and poverty and yearning and bitterness" (207).

Rather than end with that tragic realization, Burdette adds an account of his return to the Spanish Fort battle site years later, during which he discovers that every trace of the fort has disappeared. It is not "ours" after all. It belongs to the past, to "the soldier's memory" (210), and he is left to come to terms with it. Thus, Burdette emphasizes that for veterans like himself the war was not an objective matter, not simply history. It was inherently subjective, a pattern of meanings in the mind.

Of course, *The Drums of the 47th* is also a return—to the scene of battle and the testing of men, and to the maturation of a youth, a kind of military everyman, who went through it all. The book is not so much a historical report of what happened as it is a series of reflective essays on the spiritual impact of the war, an account of one man's assimilation of his Civil War experience. The reader is always conscious of the older, wiser Burdette, who reflects on his military service as he reconstructs selected aspects of it. The book is uneven in quality, as most memoirs are, but its finest sections are remarkably effective. Despite its limited usefulness for historians, *The Drums of the 47th* is one of the best-written nonfictional works based on the experience of a common soldier in the Civil War.

INTRODUCTION

Notes

1. *The Drums of the 47th* has been briefly discussed and praised in a book chapter and an article by me. See "Bob Burdette's *The Drums of the 47th*," in *Western Illinois Heritage* (Macomb: Illinois Heritage Press, 1983), 121–23, and "Civil War Accounts as Literature: Illinois Letters, Diaries, and Personal Narratives," *Western Illinois Regional Studies* 13 (Spring 1990): 58–59. Selections from *The Drums of the 47th* have also been reprinted in my anthology, *Illinois Literature: The Nineteenth Century* (Macomb: Illinois Heritage Press, 1986), 141–44.

2. Clara Burdette's *Robert J. Burdette: His Message* (Philadelphia: John C. Winston, 1922) is the source for most of the biographical information included here, but see also the biographical sketches of Burdette in the *National Cyclopaedia of American Biography* 24:356–57; *American Authors 1600–1900*, ed. Stanley J. Kunitz and Howard Haycraft (New York: H. W. Wilson, 1938), 118; and *Encyclopedia of American Humorists*, ed. Steven H. Gale (New York: Garland, 1988), 67–69. Gale also provides a fine brief commentary on Burdette's humorous writings. A helpful list of his publications is found in Jacob Blanck's *Bibliography of American Literature* (New Haven, Conn.: Yale University Press, 1955), 1:400–412.

3. For the brief mention of Burdette as a soldier, see Bryner, *Bugle Echoes: The Story of the Illinois 47th* (Springfield, Ill.: Phillips Brothers, 1905), 44.

4. "Robert J. Burdette Writes a Letter," *Sunday School Times*, Oct. 14, 1911, p. 507. All quotations from Burdette's plans for the series appear on this page.

5. "The Colonel," *Sunday School Times*, June 29, 1912, p. 410.

6. "Laurel and Cypress," in *Chimes from a Jester's Bells* (Indianapolis: Bowen-Merrill, 1897), 207.

THE DRUMS OF THE 47th

A FOREWORD

Historians prepare themselves for their tasks by much reading and by the study of great events that have marked the progress of the world's activities; writers of Philosophy deduce from the wisdom of the ages certain systems or theories; the Poets of the world drift idly on until the Muses bestow their rhythmic inspiration.

The author of this volume acquired his preparation when he himself was a maker of history. The facts told here are not compiled from other men's records, but are released from an unfading gallery of mental pictures made at a time when young life was most impressionable and the flash-light of events most unerring.

His philosophy, which has unconsciously woven itself into every thought and written page, was born of a sympathetic knowledge of human needs, activities and frailties gathered through a lifetime of loving his fellow men.

The prose-poetry of his style needed the help of no mythical Muse. It was his inheritance, as it has been his life, to drink deeply at the spiritual fountain that constantly freshened his very soul with rhythm and song. And he of all men could never lose the swing and cadence that came into life between the years of eighteen and twenty-one with the throb and the roll of the drum that, for the upholding of the great principle of life, led him to possible death.

This slender volume is offered to all who have fought in the wars of the world, that its vivid pictures may call to memory the terrible though splendid past. To all who have fought, or are fighting, the personal battles of life, that its sweet philosophy may help win them the final struggle. To all who sorrow, that its good cheer may make it possible for them to sing a new song and to know that the final beat of the Drums must be a joyous note of "Peace on earth and good will to men."

—Clara B. Burdette.

THE DRUMS OF THE 47TH

I

THE LURE OF THE DRUM

I WAS eighteen years old that thirtieth of July. I was lying in the shade of a cherry-tree, and at a window near by my mother was sewing. She sang as she sewed, in a sweet fashion that women have,—singing, rocking, thinking, dreaming; the swaying sewing-chair weaving all these occupations together in a reverie-pattern that is half real, half vision. She was singing sweet old songs that I had heard her sing ever since I was a baby,—songs of love, and home, and peace; a song of the robin, and the carrier dove, and one little French song of which I was very fond, *Jeannette and Jeannot.* It was a French girl singing to her lover who had been conscripted, and was bidding her good-by as he went away to join his regiment. The last stanza lingers in my memory:

"Oh, if I were king of France, or, still better, Pope
of Rome,
I'd have no fighting men abroad, no weeping maids
at home;
All the world should be at peace, and if kings would
show their might,
I'd have them that make the quarrels be the only
ones to fight."

Sixty years ago I first heard my mother sing that simple little song, and I have never, in all the councils of the Wise and the Great, heard a better solution of the problem of peace and war. Put a corporal on the throne, send the soldiers to Parliament and Congress, and the legislators and kings to war, and battles would automatically cease throughout all the world. It seems to me it has been a long time since a king was hurt in a fight. Nobody wears so many brilliant uniforms and such a medley of decorations as a monarch. And nobody keeps farther away from the firing line.

THE LURE OF THE DRUM

When the First Gun Sounded

It was such a quiet, dreamy, peaceful July afternoon. There was the sound of a gentle wind in the top of the cherry-tree, softly carrying an eolian accompaniment to my mother's singing. Once a robin called. A bush of "old-fashioned roses" perfumed the breath of the song. A cricket chirped in the grass.

Boom! A siege-gun fired away off down in Charleston, and a shell burst above Fort Sumter, wreathing an angry halo about the most beautiful flag the sunshine ever kissed. From ocean to ocean the land quivered as with the shock of an earthquake. Far away, from the ramparts of Sumter, a bugle shrilled across the states as though it were the voice of the trumpet of the angel calling the sheeted dead to rise. And close at hand the flam, flam, flam of a drum broke into wild thrill of the long roll,—the fierce snarl of the dogs of war, awakened by that signal shot from Beauregard's batteries.

I leaped to my feet, seized my cap and ran to

the window to wind my arms around my mother's neck.

"Mother," I said, "I'm going!"

Her beautiful face turned white. She held me close to her heart a long, silent, praying time. Then she held me off and kissed me—a kiss so tender that it rests upon my lips to-day—and said:

"God bless my boy!"

And with my mother's blessing I hurried down to the recruiting station, and soon I marched away with a column of men and boys, still keeping step to the drum.

But in the long years when the drum and bugle made my only music, often I could hear the sob, sob that broke from her heart when she bade me good-by, mingling with the harsh flam, flam of the drum that led me from her side. And at other times, when the bugles sang high and clear, sounding the charge above the roar and crash of musketry and batteries, even then, sometimes, I could hear "Jeannette" still softly singing, "All the world should be at peace." When the storm of battle-passions lulled a little at times, there would come stealing into the drifting clouds of acrid powder-smoke sweet strains of the old songs, the tender, old-fashioned melodies about home, and love, and peace, and the robin, and the carrier-dove.

I could see the window where she sat and sewed and sang on my birthday. I knew the song, and I could see how gently she rocked, and could hear how soft and low the voice fell at times. I knew that once in a while the sewing would fall from her hands, and they would lie clasped in her lap, while the song ceased as it turned into a prayer. And I knew for whom she was praying.

All the way from Peoria to Corinth, from Corinth to Vicksburg, up the Red River country, down to Mobile and Fort Blakely, and back to Tupelo and Selma, the voice and the song and the prayer followed me, and at last led me back home.

I learned then, though I did not know it nearly so well as I do now, that there is no place on earth where a boy can get so far away from his mother that her song and her prayer and her love will not follow him. There is only one love that will follow him farther; that has sweeter patience to seek him; that has surer wisdom to find him; that is mightier to save him and bring him back to home, and love and peace. What a Love that is which will endure longer and suffer more and do more than hers! What a love!

I once heard a man say,—he had never been a soldier,—"If a woman is ever given the ballot, like a man, she should be compelled to shoulder a musket and go to war, like the men."

Such a foolish, cowardly, brutal thing to say! Sometimes the government has to conscript men to make them fight for their country. When has woman ever shrunk from going to war? "She risked her life when the soldier was born." She wound her arms around him through all the years of his helplessness. Night after night, when fell disease fought for the little soldier's tender life, she robbed her aching eyes of sleep, a faithful sentinel over his cradle. She nourished him on her own life, a fountain drawn from her mother-breasts.

She stood guard over him, keeping all the house quiet when he would sleep in the noisy daytime. She stood on the firing line, battling with the foes of uncleanness, contagion, sudden heat and biting cold, protecting her little soldier in the clean sweet fortress of his home. She taught him his first cooing words that some day he might have a mighty voice and brave words of defiance to shout against his country's foes. She taught him his first step —such a wavering, uncertain little step—that some day he could keep step to the drum-beat and march with the men—a free swinging stride—as they followed the flag. She trained him up to be a manly man, to hate a lie and despise a mean action, to be noble and chivalrous. She builded a strong man out of her woman's soul.

The Woman's Harvest

And then one day, when the bugles shrilled and the drum beat, she kissed him and sent him forth at the wheels of the guns—her beautiful boy—to be food for the fire-breathing maw of the black-lipped cannon! Her boy! Heart of her heart! Life of her life! Love of her soul!

The exultant news flashes over the wires. "Glorious victory," shout the papers in crimson headlines, "ten thousand killed!"

And in the long list there is only one name she can read. It stands out black as a pall upon the white paper—characters of night against the morning sunshine—the name she gave her first-born.

And that is the end of it all. All the years of tender nursing; of tireless care; of patient training; of loving teaching; of sweet companionship; and of all the little walks and talks; the tender confidences of mother and son; the budding days; the blossoming years—this is the harvest. This is war.

When was there a generation since boys were born that women did not go to war? Never a bayonet lunged into the breast of the soldier that had not already cooled its hot wrath in the heart of a mother. While the soldier has fought through one battle, the mother has wandered over a score of slaughter fields, looking for his mangled body.

He sings and plays, the rough games of out-of-door men, in camp for a month, and then goes out to fight one skirmish. But every day and night of the thirty the mother has waked through a hundred alarms that never were. She has watched on the lonely picket post. She has paced the sentry beat before his tent. She has prayed beside him while he slept. The throbs of her heart have been the beads of her rosary.

What does a soldier know about war?

I went into the army a light-hearted boy, with a face as smooth as a girl's and hair as brown as my beautiful mother's. I fought through more than a score of battles and romped through more than a hundred frolics. I had the rollicking time of my life and came home stronger than an athlete, with robust health builded to last the rest of my life. And my mother, her brown hair silvered with the days of my soldiering, held me in her arms and counted the years of her longing and watching with kisses. When she lifted her dear face I saw the story of my marches and battles written there in lines of anguish. If a mother should write her story of the war, she would pluck a white hair from her temple, and dip the living stylus into the chalice of her tears, to write the diary of the days upon her heart.

What does a soldier know about war?

"*Five Feet Three*"

When I went into the recruiting office, two lieutenants of the Forty-seventh Illinois Regiment, Samuel A. L. Law of C Company and Frank Biser of B, looked at me without the slightest emotion of interest. When I told them what I wanted, they smiled, and Lieutenant Biser shook his head. But Lieutenant Law spoke encouragingly, and pointed to the standard of military height, a pine stick standing out from the wall in rigid uncompromising insistence, five feet three inches from the floor. As I walked toward it I could see it slide up, until it seemed to lift itself seven feet above my ambitious head. If I could have kept up the stretching strain I put on every longitudinal muscle in my body in that minute of fate, I would have been as tall as Abraham Lincoln by the close of the war. As it was, when I stepped under that Rhadamanthine rod, I felt my scalp-lock, which was very likely standing on end with apprehension, brush lightly against it. The officers laughed, and one of them dictated to the sergeant-clerk:

"Five feet three."

My heart beat calmly once more, and I shrank back to my normal five feet two and seven-eighths plus. That was nearly fifty years ago, and taking all the thought I could to add to my stature, I have only passed that tantalizing standard an inch and a half. I received certain instructions concerning my reporting at the office daily, and as I passed out I heard the sergeant say, "That child will serve most of his time in the hospital." And in three years' service I never saw the inside of a hospital save on such occasions as I was detailed to nurse the grown men; I never lost one day off duty on account of sickness. There were times when I was so dead tired, and worn out, and faint with hunger, that my legs wabbled as I walked, and my eyes were so dry and hot with lack of sleep, that I would have given a month's pay for floor space in Andersonville prison. But whenever I turned my eyes longingly toward the roadside, passing a good place to "drop out," I could hear that big sergeant's pitying sneer, and I braced up and offered to carry my file-leader's knapsack for a mile or two.

Sometimes, my boy, the best encouragement in the world is a little timely disparagement. As a rule, I am very apt to pat aspiring youth on the back, and "root and boost" with both lungs. But once in a while a good savage kick on the shins, given with all the fierceness of true friendship, puts the spring in a man's heels and the ginger in his punch to beat all the petting in the nursery.

II

THE REAL THING

I JOINED my regiment at Corinth, Mississippi. I never dreamed, when first I looked upon it in the field, how proud I was going to be of it. It was only another of the disillusions that illumined the understanding of the recruit, and showed him the difference between tinsel and gold. It taught me the distinction between dress parade and a skirmish line. For the regiment had fought at Iuka, and then marched day and night to reach Corinth in time to meet Generals Price and Van Dorn for a three-days' try-out. It was forced marching, and the barber, the manicure, hairdresser, and chiropodist had been left behind with the pastry cook—back in Illinois. My regiment! In my dreams it had always looked like a replica of the Old Guard at Marengo. Now it looked more like the retreat from Moscow. Save that it never retreated.

Uniforms grimed with the dust of the summer

roads and the rains and mud of the spring campaigns. Some of the soldiers wore military caps, but none so new and bright and blue and bebraided as my own. Hats were largely the wear. The army hat of "the sixties." A thing fearful and wonderful when it was new, with a cord that was strong enough to bind an enemy hand and foot, and terminating in tassels big enough and hard enough to brain him. One side looped up with a brass eagle, not quite life-size. The inflexible material of the hat made it break where the side was turned up. The crown was high and the brim was flat, the general effect being a cone with a cornice. Sometimes the soldier creased a pleat in the top, that it might resemble the Burnside hat, by which name, indeed, I think it was called. This broke it in two, and let in the rain.

Well, my comrades had marched in this grotesque head-gear in the dust and in the rain. They had fought in it. They had slept in it. They had used it for a pillow in the resting halts on the march. On occasions they had carried water in it. One warrior told me he had boiled eggs in his. But you can't tell. You may guess what it looked like when I first saw it. I can't. I saw it, and I couldn't recall anything I had ever seen in my life that it faintly resembled.

The most comfortable way of wearing the trousers on march was by tucking them into the legs of the army sock. Oh, yes; plenty of room. A man could put both legs into one army sock of the sixties. I never tried slipping one over an expanded umbrella. But that was only because there was no umbrella. Wearing the sock over the legs of the trousers was the best, and save in the new days of the sock, the only way to hold it up. The sock was made by machinery. In one straight tube, I think, and then pressed into sockly shape. This lasted until they were washed the first time. Then the article reverted to type, and became the knitted tube from which it had evoluted.

Recollections of the Old Army Shoe

The shoes were not dancing pumps. But of all

the things that ever went on a man's pedals, the old army shoe was the easiest, the most comfortable and comforting thing that ever caressed a tired foot. I think among half a hundred recruits with whom I went to the regiment there were at least twenty-five pairs of leg boots; well-fitting boots; made by good shoemakers at home, and costing good money. After the first long march possibly half a dozen pairs survived intact.

And they lasted only until we could draw the government common-sense shoe. Affection could not make that shoe beautiful. But prejudice could not make it uncomfortable. If you put it on side-wise, it would not "run over." Its process of wearing out was peculiar. A few days before dissolution the shoe displayed symptoms of easy uneasiness. It flattened out a little more across the toes, which was impossible. Always easy, from the first day it was worn, it grew easier day by day until it suddenly became luxurious to effeminacy. Then, on a long muddy hill-climb of Mississippi clay, the sole pulled off back to the heel, the upper spread itself like a tanned bat, and the shoe was gone. That was all. The soldier swore his astonishment and disgust, girdled his shoe with strings, and wore it sandal fashion until he could draw a new pair from the quartermaster, or procure a pair from one of the many sources of supply which were an open mystery to the quartermaster's department and matters of profound surprise to the innocent soldier, grieved at being wrongfully accused of "conveyance."

My dusty, war-worn, weather-beaten, battle-stained regiment! About four hundred men. Was this war? Were these "soldiers"?

Then I watched the companies march out to dress parade. It wasn't drill-room marching, and there was no music to time their steps. But it was the perfection of walking. The men swung along with a free stride learned by natural methods in muddy roads, on dusty turnpikes, on steep and winding trails that climbed from the plain to the hill-top. They kept step without knowing it. They marched the best way because it was the easiest way. Then the line of the parade. From company to company officers and sergeants barked a few terse orders; a little shuffling of feet, and the line stood petrified at attention.

Just to be a Recruit!

An engineer couldn't have altered it to its betterment. Then the adjutant barked "Front!" and the parade was formed. Square shoulders, full chests, breathing deep, and slow, and regular as a race-horse; easy poise of body, hands resting on the ordered muskets lightly as they would hold a watch or a pencil, yet so firmly that when the command, "'Der—hmm!" came, every piece swung to a "shoulder" like the movement of a machine. Through the old-fashioned manual of arms, unintelligible to the soldier of 1914, even as was the "Scott manual" to the men drilled in "Hardee," there was the same precision of movement; the click of the hands in one time and two motions, varied by the order, as the piece moved from the old to the new position, or fell with a simultaneous thump on the turf to the "order." If a man came through out of time, the discord was heard the entire length of the line, and the eyes of the colonel went to the face of the laggard like bullets, while the nearest sergeant growled sweetly through his mustache at the culprit. Mustaches were worn in the army of the sixties; every face had one. Not an eye in the line looked toward another man for a lead. Every eye straight to the front, and every man save the nervous recruits knowing just as well as the colonel the order of the manual on parade.

The "troop, beat off"; the band marched down the line to slow music, and countermarched back at quick time—*Rocky Road to Dublin, The Girl I Left Behind Me,* or the everywhere popular *Garry Owen,* or some lively air to which the regiment had words of its own, *The Death of My Poor Children* being a favorite of ours. The first sergeants took command of their respective companies and marched them back to quarters, and my heart thrilled to watch them, while I wondered, as I vainly tried to imitate them, if I would ever learn to walk like that! Now, the uniforms seemed to fit like dress suits; the hats were jaunty as the caps; the accouterments were ornaments; every joint in the soldier's body was "ball-bearing,"—play of the hips, swinging arm, the heel-and-toe walk, twenty-miles-a-day gait,—all the dancing schools in America couldn't put the ease and grace into that soldier's poise and movement that months of marching had done. How proud I was just to be a recruit in such a regiment!

A greyhound looks prettier than a bulldog. That's because it's built for running. But a bulldog is built for fighting. That's why you always turn to look at a bulldog when you pass him in the street. As you turn to look you smile at a stranger who has turned at the same time. The stranger nods his head. You are both thinking the same thing. That's the way you feel when, after witnessing a prize drill of the East Haddam Invincibles, uniform dark and sky-blue, picked out with white; gold stripes down the trousers; frogs across the breast of the coat; red, white and blue plumes in the caps; white gloves; buttons by the gross; patent leather knapsacks quite as large as a bon-bon box, you suddenly meet a regiment coming back from the war. It's like coming out of a heated stuffy ballroom, sickly with perfume, to feel the keen north wind of November blow into your face with the breath of a new life—strong, exultant, thrilling.

The Drums of the Forty-Seventh

We had a brass band when we went to war. But when the regiment got to the front it traded the brass band for a fife and drum corps. Because the regiment is a fighting machine. Doesn't the band go into battle? Sure. Not to nerve our fighting courage with spirit-stirring strains of stormy music. The musicians tied simple bandages of white or red around the left arm, and reported to the surgeon for duty. They sought out the wounded and carried them back to the field hospital, sheltered behind some merciful hill, under the tender shadow of a clump of trees. They found the dead, and carried the poor sacrifices to the rear to lay them in silent ranks for their last bivouac.

Some of their human burdens wore the uniform we loved. And some of them were clad in the gray against which we fought. But the blood-stains, like cleansing fountains, washed out all hate and malice. War rages over the embattled field, a storm of passion. Under the trees in the rear of the fighting lines, when bullet and bayonet and shell have wrought their hurt, soft Pity moves, a ministering angel of God's sweet compassion. Her healing hand touches with equal tenderness the wounds of friend and foe. And we were friends. We were brethren a little while estranged. And Love is strong as death and stronger than hate. And truth outlasts all misunderstanding.

We had a "fighting band." Our musicians unslung their drums when the last mile was growing longer than a league, and carried us into camp with *Jaybird, Jaybird,* shouting fresh. In the morning it played us out of camp with *Garry Owen.* And when the skirmishers deployed, the musicians piled their drums back near the baggage and lightly trotted in open formation close up to the firing line, with extra canteens and ready stretchers and emergency bandages, and much cheery chaff. These preparations looked chillingly in earnest. For it was always very dangerous to go into battle, especially, as one irreverent private remarked as he looked around on the unsheltered plain, "Hard lines for us, boys; there aren't half enough trees for the officers!" One of our drummers—the youngest—was a tonic for a faint heart. Johnny Grove; he could drum to beat a hail-storm on a tin roof, and he had a heart full of merriment and a tongue as ready as a firecracker. Death came very near to him many times, but he always laughed when he heard the boy, and passed on, and Johnny still lived with a heart as mellow as then it was light, until a few years ago.

The drums of the Forty-seventh—they time a quicker throb to my old heart now, when I think I hear them again, on a rough road and a steep grade. The drummers are old men; old as myself. And again they are playing the regiment into camp. The fifes blow softly as flutes. The roll of the muffled drums, tender as the patter of rain on autumn leaves, times the slow steps of old soldiers with the *Dead March* to which we listened so oft when life was in the spring-time.

> "There's nae sorrow there, John;
> There's neither cauld nor care, John,
> The day is aye fair
> I' the Land o' the Leal."

III

READY TO KILL AND TO LAUGH

A SOLDIER expects to see somebody killed. A battle-field is a very dangerous place. Even a skirmish-line in a little reconnaissance is an unsafe locality for a picnic. There is always more or less—and usually more—peril of exposure during the storms of war. That brilliant Georgian, Henry W. Grady, of Atlanta, made a very striking criticism of General Sherman when he said that "he was a great general, but he was mighty careless with fire." When the recruit is handed his Springfield rifle, and the corporal shows him how to load it, and teaches him the best method of taking careful and accurate aim, and how to secure the most rapid firing with most effective results, the soldier is aware that he is not going to fire blank cartridges. In fact, he is given no blanks. I never saw one all the time I was a soldier except when they were dealt out to the firing squad at a funeral, when the man over whom we were to fire was already dead. The weight of the forty rounds in the cartridge-box assured the soldier that he was carrying forty bullets, every one of them capable of killing any living thing it hit—Texas steer, grizzly bear or man.

And the recruit understood very plainly that every bullet was meant to have its billet in a human body. It was made to kill some human being, and he was appointed by the government as its active agent to direct the bullet to a vital spot in the right man. He was especially warned against the unsoldierly sin of firing too high. That is the common fault, even of the old soldier. Toward the end of a long day of fighting, when the whole body is wearied, and the left arm is especially tired with the weight, and the right shoulder is sore and sensitive with the kicking of the musket, the weapon pulls the arm down, and the bullet, spiteful but harmless, kicks up the dust only a little way in front of the soldier. But as a rule he shoots too high. The repeated expostulation of the sergeants is, "Fire low, boys; fire low! Rake 'em! Shin 'em!" Otherwise the recruit will not kill anybody, and that is what he is shooting for. It is what he is paid for. That is his "business."

Trained for the Business of Killing

It sounds very cold-blooded, but it is all cold fact. Killing is the object of war. "You can't make omelettes," said Napoleon, "without breaking eggs." When you knew him at home, the recruit was one of the happiest, best-natured boys in town: kind-hearted, sympathetic, gentle as a girl. But now that he has enlisted, and has a gun, and is daily taught and trained how to load and fire with deadly aim, it is his duty to kill as many men as he can, before the one who has been detailed for that purpose by an officer on the other side kills him. The glittering bayonet which the soldier is taught how to "fix" on the end of his musket is not an ornament. It is made sharp at one end, a wicked sort of triangular bodkin, so that a vigorous lunge will drive it into the breast of a man up to the muzzle of the musket. There is nothing strictly ornamental about a rifle. In case the bayonet should break, and all the cartridges are burned, the soldier is taught how to make a most effective deadly weapon out of the butt of his musket. Everything about a military equipment is dangerous to human life. Even the rations are often condemned, especially embalmed beef.

You might recognize a recruit in an old regiment by the careless manner in which he handles his piece. He leans it up against a tree without seeing to it that it is firmly balanced and braced in its place. When he stacked arms, it was an old soldier who tested the stability of the "stack" with a little shake. The recruit carried his gun any which way on the march, until he was taught better by gentle caution, stern reprimand and jarring kick. The old soldier knew that a musket was dangerous "without lock, stock or barrel."

One night in 1863 we bivouacked in an old Confederate camp ground which had been hastily evacuated on our uninvited approach. We found the rude bunks very comfortable, and not more over-crowded with inhabitants than the abandoned bunks of an old camp are liable to be, without regard to previous political affiliations.

Next morning we broke camp to go on in pursuit of our retiring hosts. The "assembly" had sounded, and while we were lounging about waiting to hear "Fall in," a soldier in my own regiment found an old revolver under one of the bunks. It was one of those antique, self-cocking curiosities known as an Allen's "pepper-box," the most erratic and unreliable weapon of death ever designed to miss anything at which it was pointed. It was commonly supposed that a man couldn't hit a flock of barns with one, if he were standing inside the middle barn. But this time the soldier, knowing the character of the "pepper-box" for general inaccuracy, pointed his find at a comrade, cried, "Surrender, or you're a dead reb!" and pulled the trigger. The deadly accident discharged a load it had probably carried ever since the war began. The living target fell on his face with the blood streaming from his mouth. He was shot through the lungs, and died in a few minutes. And unanimously the man who shot him was condemned with savage harshness, unmollified by one word of pity, sympathy or excuse.

"What do you think a revolver is," demanded one of his own company, "a watch-charm?"

Soldiers do not "play" at war. They do not use their arms and accouterments as playthings. They do not care to "play" soldier. One evening after a long tramp through a hilly country in Mississippi we went into camp in the heart of a nest of hills, rugged and ragged, and dense with woods and undergrowth and tangling vines, with water hard to get at, and were informed that we would remain in camp in that unpromising land for about three weeks.

Such a shout of joy, loud, long-sustained and oft-repeated, as went up from three hundred tired men! Why? There was no drill ground. No wide stretches of level fields; no broad valleys where we could practise methods of approaching and crossing a shallow creek, easy to ford, but just as wet to the feet as the Pacific Ocean. The soldier did not love to drill. On the other hand, his colonel was perfectly infatuated with the game of war. And brigade drill! This is the delight of the general. It is a movement in masses. The regiment is the unit. It is an inspiring spectacle to mounted officers. To the infantryman, down in the dust and stubble of old cotton and corn-fields, seeing nothing but the monotonous wheels and half-wheels, rights and lefts into line, facings and halts, it is indescribably dull and tiresome. Occasionally, when after some complicated movements he finds himself and the regiment in the middle of the many-acred field, perfectly formed in the most beautiful hollow square a brigadier ever smiled down upon, there does come a thrill of pride and delight into his soldierly heart, for here is something he can appreciate. He brags about it more than his general. But, as a rule, he classes drill with hard work. At least he says he does.

Laughter Shakes the Line of March

But if a soldier grumbles at many things, he laughs at anything, and many times just as heartily he laughs at nothing. "For once, upon a raw and gusty day" as ever "the troubled Tiber chafed," we were marching through a pelting rain, splashing through the slushy mud. A tired soldier sought an easier footway up on the sloping roadside, and pulled off both shoes, one after the other, in the sticky clay bank,—an insult in the face of misery. The men who saw him roared with pitiless mirth. The next company, which could not see, howled in sympathy with laughter they could not understand. Down the line it went, increasing in volume as it got farther away from the cause of the unkindly merriment. The regimental teamsters caught it up, and their stentorian haw, haw, haws set the mules to braying. This passed on to the Second Iowa Battery, and the gunners made the soaking welkin ring with their cachinnation. It drifted back to the Eighth Wisconsin, and they slapped the spray out of their soaking trousers as they added gesticulation to emphasize their guf-faws. And all along the column the straggling groups of happy freedmen shrieked with ignorant delight after the manner of their mirthful kind.

Well, that's one mission of laughter. Every soldier will tell you of such things. There is one army story that echoes from the Potomac to the Mississippi. Vociferous cheering, "a cry as though the Volscians were coming o'er the wall," breaks out at some point in the marching column. It goes down the line of march, a great wave of laughter, cheering, exultant; increasing in jubilation until it reaches the rear-guard in a mighty climax of re-joicing uproar, that would terrify the troopers of the enemy hovering on our rear, but that they un-derstand it all as well as we, for the custom was as one in both Union and Confederate armies. Either a favorite general has galloped down the column, or a frightened rabbit has dashed across the line of march. In either case it is the same—the bravest of the brave, the fightingest general known of that division, or the timidest, scared-to-deathiest little animal in the world has received the same meed of tumultuous applause. Every veteran will tell you that his regiment, the fighting Hundred and Onety-Oncest, had an exclusive saying on such occasions: "Old Smith, Mower, Sherman, Sheridan, Hubbard, Logan," etc., etc., "—or a rabbit!"

Camping Stories Ancient and Modern

There were few copyright stories in the army. A California regiment crossing the plains to join the army east of the Rockies would meet its own anecdotes, told with a nasal twang by the Steenth Vermont. Army stories are uniform as army rations. The soldier on the stormy march who longed to be under the old barn at home, because it would be so easy to get into the house; the one who asked the sutler if his pies were sewed or pegged; the one who, when the dear old lady listening to his account of the battle asked him why he didn't get behind a tree, scornfully replied that there weren't half enough trees for the officers; the soldier who was surprised on picket by his brigade commander, with his gun taken apart, oiling it, said, "You just wait till I sort o' git this gun sort o' stuck together and I'll give you a sort o' salute," was a Confederate, but we stole his story just the same; the soldier, missing everything at target practise, asked by his impatient sergeant where under the sun his shots went, who replied, "They leave here all right; I can't tell where they go after they get away from me"; the sentry who challenged, "If you don't say Vicksburg mighty quick I'll blow your head off"; the Irishman who said "Bags" when the countersign was "Saxe"; the slovenly soldier who, reprimanded on inspection by his captain, "How long do you wear a shirt?" replied, "Thirty-four inches"; the jayhawker who killed a sheep in self-defense because it ran after him and tried to bite him; all these narratives and many more were ascribed to men in my own regiment. Quick as we heard a new story we found the hero of it, in our own ranks. All the regiments in both armies follow the same patriotic custom. The wrathful shout of Frederick the Great to his recoiling grenadier, "What, then, do you want to live forever?" is repeated of every colonel since his day; we told it of five of ours. For when we exchanged stories with our prisoners, hoping to get some new anecdote material for our regimental fame, lo, the captives of our bow and spear told us our own threadbare tales about the Eighth Georgia and the Louisiana Tigers. Doubtless the guards at Libby Prison suffered the same bitterness of disappointment when they sought to add to their own stock of "the best and latest." The army stories with which the archers of Parthia and the left-handed slingers of Benjamin were wont to set the tables in a roar were easily adapted to the stage settings of the time by the musketeers of Frederick and the Grenadiers of the Old Guard. And now the pontoon stories are the uncopyrighted property of the aeroplanes and dirigibles.

IV

STORMY WEATHER CHRISTIANS

MAY 14, 1863, and a rainy morning at Mississippi Springs. The bugles sang reveille as sweetly as though the sun was shining on the drenched violets by the muddy roadside and in the dripping woods. The drums beat sullenly, for like many more delicate musicians they are very sensitive to changes of the weather, and never like to get their heads wet. It takes all the thrilling "rat-a-plan" out of their chest notes, and makes their voices flat and tuneless as they thump out "Three Camps," "Slow Scotch," their double drags and three rolls.

But the bugles!

Their voices never change. I have heard them in the midst of the storm of war on a blood-drenched battle-field come ringing down the broken lines, breaking through the pungent powder smoke, their voices of command clear as the song of a meadow-lark calling through a bank of fog or a cloud of drifting mist. Strangely sweet, the bugle call in the midst of the battle clamor—the roar of the guns, the fierce rattle of musketry, "the thunder of the captains and the shouting." Heart-breakingly sweet. The soldier starts sometimes as though he heard the echo of his mother's voice calling him out of the passion of carnage, calling him back to her side—back to her arms, back to her tender caresses, soothing the storm of battle rage in his young heart,—calling him to home and peace, with the old love songs, the cooing dove and the whistling robin.

Then the bugle, sweetly as ever, calls yet more insistently, and a great thundering shout from the colonel drowns the mother-voice—"Fix bayonets! Forward—guide center—double quick—follow me, boys!" And the wave of the charge carries the line forward on a billow of cheers in a tempest of fighting madness. And still the bugle calls, just as sweetly and just as insistently, as though a beautiful queen were urging her soldiers on to glory and victory—Deborah singing *The Charge:*

How can anything so beautiful set a man on to fight and kill? Well, it does. A soldier in a fatigue uniform looks like a dude alongside of a civilian in his fishing clothes. There is good music in the beer halls; better, sometimes, than you can hear in your home church. A regiment marching down street behind its military band Sunday morning is far more alluring in appearance than the throngs of worshipers straggling along to worship. Why is a battle-ship more attractive than a ferry-boat?

The Lure of the Fighting Spirit

If you are walking with a friend, and pass an old man, white-haired, face lined with furrows of time and thought and toil, stoop-shouldered, leaning heavily on his cane as he steadies his steps, and the friend says to you, "That is Doctor Soulsaver; he has been pastor of the same church in this city fifty-two years, and his people won't let him resign."

You say "Uh-huh!" glance around at the old man as he totters by, and go on talking. But if you meet a man with his civilian suit cut in military fashion, a white mustache ornamenting a bronzed face, swinging his cane to show that he carries it as a plaything, and your companion says:

"That's General Smasher; he's been in the army since he was a boy; been in more battles than any man living; been wounded ten times; the hardest fighter in the American army; never was whipped."

You stop and look after the old mustache until you forget what you had been talking about. You'd like to meet that man. Why didn't you run after the old preacher and shake hands with him? You're a church member. Why did you feel more interest in the old soldier? You tell.

Why isn't virtue as alluring to the senses as evil? Better is wisdom than folly; sweeter, purer, nobler, lovelier. Yet it is written, "When we shall see Him, there is no beauty that we should desire Him."

Sunny mornings or rainy mornings, the bugles sang as cheerily as so many meadow-larks, the bird with never a plaintive note in his song, whether the wind blow from the south with perfume, north with biting cold, or east with fog and rain, or west with a roaring cyclone. And this morning the bugles called out of the soaking chrysalides of the blankets a lot of crowing soldiers who echoed the bugles in their own music. A soldier's dreams must be sweet, for always—so nearly always the exceptions are not worth noting—he wakes up in high good humor. Such good medicine is sleep. And he sings the reveille with the bugles—

REVEILLE

The day-star shines upon the hill,
  The valley in the shadows sleep;
In wood and thicket, dark and still,
  My comrades lie in slumber deep;
Far in the east a phantom gray
  Steals slowly up the night's black pall,
And, herald of the coming day,
  Softly the distant bugles call—

THE DRUMS OF THE 47TH

"I can't get 'em up,
I can't get 'em up,
I can't get 'em up in the morning!
I can't get 'em up,
I can't get 'em up,
I can't get 'em up at all!"

A thought of motion at the sound,
As though the forest drew its breath,
And belted sleepers on the ground
Move restlessly, like life in death;
And slumberous echoes, here and there,
Awaken as the challenge floats,
And clearer on the morning air
Ring out the cheery bugle notes—

"The corp'ral's worse than the private,
The sergeant's worse than the corp'ral,
The lieut. is worse than the sergeant,
And the captain's the worst of all!"

And while the thrilling strains prolong,
  Flames into rose and gold the day,
And springing up with shout and song,
  Each soldier welcomes march or fray;
Through wooded vale, o'er wind-swept hill,
  Where camp-fires gleam and shadows fall,
Louder and sweeter, cheerily still,
  Ring out the merry bugle's call—

"I can't get 'em up,
I can't get 'em up,
I can't get 'em up in the morning!
I can't get 'em up,
I can't get 'em up,
I can't get 'em up at all!"

A cold breakfast, scalded down with boiling coffee, black as night and strong as prejudice, put the spring in our heels, and we were ready for whatever the day might bring to us. We twisted our wet blankets, a load in themselves, and looped them over our shoulders. My regiment was a marching and fighting regiment, and knapsacks were luxuries of effeminacy, indulged in only in winter quarters. The sodden drums beat a doleful accompaniment to the merry squeaking of the fifes as they whistled us out of camp into the canal-like road with *Garry Owen na gloria.* Splash, splash, splash, through the mud. We wrapped ourselves in the shelter of our rubber blankets. But the steady rain found open folds at our necks, and crept in and trickled down our backs in little zigzag trails of moisture that found its way down into our shoes. As long as a soldier can keep his feet dry, he is comparatively comfortable. But when the water begins to sqush, sqush, in his shoes, Comfort bids him a reluctant farewell, and Misery, perching heavily between his shoulders, says:

"Would you mind carrying me until it clears off?"

The warrior does "mind," but carries him just the same. His feet slip in the mud, and this makes marching hard and slow. A calvaryman, galloping down the column with an order from the front to the rear, or vice versa, splatters the infantryman from head to foot with mud and water, and is pursued for the next three miles of his career with volleys of sarcastic and abusive comments on his horsemanship, his horse, his yellow stripes, his clanking saber, his personal worthlessness and his disgraceful pedigree that make his ears tingle and his heart boil with wrath. A baggage wagon stalls on a steep hill, and the soldiers come to the rescue of the struggling mules and help them up the long muddy Hill of Difficulty, the name of which is Legion.

"What's in that wagon?" asks a recruit who has twice fallen in the mud, in his zeal to do his whole duty by the mules. "Ammunition?"

"Naw!" scornfully replies the veteran. "Suppose I'd break my back pushin' a load of ammunition? Them's hardtack."

And that's worth while, and the soldier hopes his long-eared comrades will reciprocate his help and bring that wagon into camp on time at night.

The Soldier's Rainy-Day Religion

Splash, splash, splash. The arms ache with the weariness of carrying the musket in one position, and that not the easiest one by any means. But the musket is as precious as the hardtack. The soldier may get soaked to the bone. He'll fight just as well. But that gun must be kept dry. He carries it at "secure," under his arm and under the protection of his rubber blanket. Now and then he looks at the hammer and nipple to see that they are dry. He may want to use that piece of hand artillery before night, and he cares for it like a baby. He may have to shoot somebody with it some time during the day. And suppose, when that time comes, the powder in the musket is wet. How can he carry out the decrees of fate concerning the man he is detailed to kill? Wet powder has no more place in a musket than a knot-hole in a barb-wire fence.

He has to stop wasting caps, and pick dry powder into the nipple with a pin. Tedious work it is, and the unpleasantest thing about it is that the man he was to kill may get tired of waiting and fill him full of large irregular holes by way of reproach for his dilatory tactics.

Really, the soldier grumbles less and wants to fight more, in all the discomforts and irritations of a stormy day over muddy roads in a hilly country, than he does in June weather through a pleasant land. He'd like to fight the people whose conduct has dragged him away from his happy home. But if one of his comrades loses patience and breaks forth in bitter reviling of the rain and mud and the war, he helps to smother him in an avalanche of raillery and chaff. After that the column is in a happy self-approving frame of mind for several miles. The worst environments bring out the best in the soldier. He braces himself to meet adversity, as he would meet any other enemy. He prides himself on being above the demoralizing influences that break down weak men. He may swear a little, which is more than enough; and he may drink too much, which is when he drinks at all. And he kills a few people. Which is his first and constant duty. What's a soldier for? But he believes in rainy-day religion. His standard of manhood is high, and he found it in the Book his mother gave him: "If thou faint in the day of adversity, thy faith is small."

V

THE MURDER

THEY killed him on the early afternoon of a May day, May 14, 1863. We never found the man who did the deed, although there was no pretense of concealment about it. It was committed in broad daylight, in the presence of hundreds of men. The murder was officially reported, but there was no investigation. Indeed, it was not called a "murder" at all. It was simply reported as a "casualty." "Casualty"—"what happens by chance," the dictionary says; "an unfortunate accident, especially one resulting in bodily injury or death; specifically, disability or loss of life in military service." It is something to be expected. It is taken for granted. But the man himself, who made the accusation as he was dying, called it "murder."

A dull staccato thunder of guns in the distant

front, a galloping staff-orderly giving an order to Colonel Cromwell, which he shouted to us; a sudden barking of many commands from the line officers; a double-quicking of the column into the line, and almost in the time I have written it we were in line of battle in the woods before Jackson, Mississippi. I heard Captain Frank Biser shouting his customary "instructions to skirmishers" as he deployed A and B Companies into the skirmish line, and they disappeared amid the scrub oaks,—"Keep up a rapid fire in the general direction of the enemy, and yell all the time!" He was very specific regarding the kind of "yelling," which was to be emphatically sulphurous. The regiment followed to the brow of the hill that looked down on the creek winding in muddy swirls and many meanderings across the level meadows. Far to our right we could hear our own battery, the Second Iowa, its bronze Napoleons throbbing like a heart of fire. And at our left the Waterhouse Battery, of Chicago, was baying like a wolf-hound at the gray battalions down by the little Pearl River. We were supporting that battery. And we were ordered to lie down and keep ourselves out of sight.

The Man Who Stumbled

This seemed to me excessive caution. I was a recruit in my first battle. I called it a battle. The old soldiers spoke of it as a fight. Whatever it was, I wanted to see it. I rose up on my knees to look about me. It didn't look like any picture of a battle I ever saw in a book. The man with whom I touched elbows at my right, Doc Worthington, of Peoria, and an old schoolfellow before we were comrades, said with a note of admiration in his voice:

"Haven't those fellows got a splendid line?"

I saw the long line of gray-jacketed skirmishers doing a beautiful skirmish drill. Puff-puff-puff the little clouds of blue smoke broke out from the gray line moving through the mist that was drifting across the field. I saw the blue-bloused skirmish line come into view from the woods at the foot of the hill. I saw a man stumble and fall on his face.

Not until he did not get up and go on with the advancing line did I realize that he had not stumbled.

I had a strange trouble with my breath for a boy with lungs like a colt and a heart that is strong unto this day. An officer came riding down the line, pulled up his horse, asked a soldier for a match, calmly lighted his pipe, puffed it into energetic action, and rode down the hill after the skirmishers. How I admired his wonderful coolness! By the time I went into the next battle I knew that the pipe trick was not a symptom of daredevil, reckless coolness, but only of natural human nervousness. The man smoked because he was too nervous not to.

I saw the skirmishers now and then rush suddenly together, rallying by fours and squads as a little troop of cavalry menaced the line with a rush,—a charge, we called it then. I saw them deploy just as quickly, and heard them cheering as a rapid volley admonished the troopers with a few empty saddles. Then I saw the gray line advance resolutely, and with much dodging and zigzagging our own skirmishers were slowly falling back to their line of support. The guns of the Waterhouse battery, fiercely augmenting their clamorous barking, suddenly fell silent. The gunners swabbed out the hot cannon and then stood at their stations.

"Why do they stop firing?" I asked.

"They are letting the guns cool," said a corporal.

"They are going to get out of this," said Worthington; "those fellows are coming up the hill."

I was looking at a young artilleryman. He was half seated on the hub of one of the Waterhouse guns, resting his face against the arm with which he cushioned the rim of the wheel. He was a boy about my own age, not over nineteen. He was tired, for serving the guns in hot action is fast work and hard work. His lips were parted with his quick breathing. He lifted his face and smiled at some remark made to him by one of the gunners, and his face was handsome in its animation—a beautiful boy.

I heard a sound such as I had never heard before, but I shuddered as I heard it,—dull and cruel and deadly. A hideous sound, fearsome and hateful.

The young artilleryman leaped to his feet, his face lifted toward the gray sky, his hands tossed above his head. He reeled, and as a comrade sprang to catch him in his arms the boy cried, his voice shrilling down the line:

"Murder, boys! Murder! Oh, murder!"

He clasped his hands over a splotch of crimson that was widening on the blue breast of his red-trimmed jacket and fell into the strong arms of the comrades who carried him to the rear. Him, or—It.

The rain began again and the warm drops fell like tears upon his white face, as though angels were weeping above him. I watched the men carry him away to where the yellow flag marked the mercy station of the field hospital.

Fear, before unfelt because unknown, clutched my heart like the hand of death, with the voice of that hissing spiteful bullet. My very soul was faint.

I did not know—I shall never know—who shot this boy. Nor, I think, does the man who killed him. Another boy, maybe. For there were as many schoolboys in the Confederate armies, it seemed to me, as men.

What Friends They Might Have Been!

Why, the war was only a year old. The boy who fired that rifle-shot—his mother's good-by kisses were yet warm on his cheeks and lips. Only yesterday his sister unwound her arms from their caressing clasp about his neck to let him go to the war. Such a warm-hearted boy he was, they would tell you. Affectionate as a girl. A loving, impulsive southern boy. From the time that he first knelt at his mother's knee and learned the prayer that all mothers, north and south, teach their boys alike, he had knelt morning and evening before the Prince of Peace and prayed that his heart might be kept pure and sweet, and gentle and kind.

And now?

See what he had done! He had committed a deed of death so far away from all his boyish thoughts that he had never prayed against it.

And the boy from the Northland whom he had shot — the other boy, who had been trying to kill him with the terrible six-pounders. Why, his mother, too, had kissed him good-by in the doorway of that far-away Illinois home, with her tears raining through her kisses, just as the rain-drops of the May shower fell upon his white face a minute ago. His sister had sobbed her good-by as she held him close against the heart that had loved him since he was her tiny baby brother—the heart that now would break for him. A quiet gentle boy, they would tell you. Always that smile on his face I had just seen. And all the years, as he knelt with bowed head and clasped hands, unknown to each other, his prayers and those of the Alabama boy had mingled as they ascended to the same heavenly Father.

What true-hearted friends they might have been,

those two boys, had they met some time other than that sunless rain-swept day in May. And yet, not half an hour ago, the boy from Illinois had been working at those murderous guns like a blacksmith at his forge. When his gun, with a fierce breath of flame roared its defiance, shook out a murky banner of blue smoke, and sent its messenger of death screaming into a group of men and boys down in the meadow, how quickly that boy from Illinois sprang with his sponge staff to wipe the black powder stains from the grim lips, and cooled the rifled throat, hot with hate and death. How proudly he patted its grim sides when it made a "good shot" —that is, when it killed somebody. And then, sitting on the hub of the wheel, the battle rage subsiding in his heart as the sullen gun cooled at his side, the longing came dreaming into his eyes, his thoughts drifted away to a home up beside Lake Michigan, his mother and sister came into his heart—

And then a boy not unlike himself, a boy who had been watching the deadly work of the Waterhouse guns, a boy standing in a little clump of bushes in their May bloom, raised his rifle, aimed carefully at the cloud of smoke drifting slowly away from the last shot of that terrible gun, and, without knowing or seeing who was sitting behind that beautiful screen, fired.

And killed a boy to whom his soul might have knitted itself, even as the soul of Jonathan clave to David.

"Murder! Oh, murder, boys! Murder!"

Well for that boy in the Southland that he could not hear that cry. And well for all our boys in all our land if they shall never hear it.

The Cry Through the Starlight

The bugles called sweetly and imperiously, the colonel's voice rang out stern, peremptory, inspiring, the line sprang to its feet, and with mighty shouting rushed forward like unleashed dogs of war. Thundering guns, rattling musketry, cheering and more cheering, a triumphant charge, a wild pursuit, a mad dash—we were over the works and into the city. That night my regiment bivouacked in the pleasant grounds of the beautiful capitol of Mississippi. My first battle, and it was a victory—a victory—a brilliant victory! And I had a soldier's part in it. How proud I was! I could not sleep. I mentally indited a dozen letters home. And again I whispered a prayer, and looked up my good-night at the stars.

Calm, silent, tranquil. Undimmed by the smoke of the guns. Unstained by the blood that had smeared the meadow daisies. Unshaken by all the tumult of charging battalions. Sweet and pure, the glittering constellations looked down upon the trampled field and the dismantled forts. Looked down upon the little world in which men lived and slept; loved and hated; fought and died. The quiet, blessed, peaceful starlight.

Far away, yet thrilling as a night alarm, came dropping down through the starlight the cry that went up from the sodden earth ages and ages ago:

"Murder! Oh, murder!"

My thoughts went northward, because I could

not sleep, to the little home in Peoria where mother and sisters waited for me. Slowly, although I tried to keep them away, my thoughts came back to the battery on the brow of the wooded hill where the purple violets smiled through the strangling smoke of the guns. With a troubled mind I thought of other mothers and sisters who waited in northern and southern homes. I laid my arm across my face to shut out something that dimmed the starlight and marred the glory of victory with the stain that marked the altar of prayer and sacrifice when the world was young and fair. I would not allow myself to think of hideous and hateful things. I would think of love and home, and the whistle of the robin, the song of the meadow-lark, and the mother voice, soft and sweet and dovelike, cooing the old love-songs.

Still, even as I slept and dreamed of a victory won and of other fields of glory and triumph to come, down through the starlight came the echo of that fainting cry under the wheels of the guns:

"Murder! Murder, boys! Oh, murder!"

VI

THE FLAG

WHEN the bugles have called a sweet "tira-lira-la" that sounds more like the refrain of an old love-song than a battle-cry, a thrilling call, a magic word, that suddenly opens the long marching column like the sticks of a colossal human fan, infantry and batteries double-quicking or galloping to right and left into the extended battle-line, there follows a halt of preparation. The panting line is quickly "dressed", and as a hurrying aide halts beside our colonel, hastily to explain to him the position of the batteries and the other infantry regiments with reference to his own command, the adjutant fires an order or two at us:

"Front!" "Or-*der*—h'arms!"

Then the colonel commands, in a tone so intense that it reaches center and flanks at once:

"Load at will—load!"

The metallic ringing of the rammers springing from their sockets; the thud—thud—thud as they drive the cartridges home; the clicks that tell the colonel the hammers are back on the caps, and the life of a man is hidden away in the breech of the rifle. Then—

"Fix—bayonets!"

Rattle and click of metal against metal all along the line.

"Carry—h'arms!"

And the regiment stands as on parade or review. At "carry," because, under the old Hardee tactics, at "carry" the musket was most readily raised to "aim" or dropped to "charge bayonets." Now we are ready for anything. A bugle calls again, sweetly as a mother might call her laughing children in from play. The colonel interprets the well-known syllables—

"Forward—guide center—h'march!"

When We Marched Without Music

The line moves forward. Not a note of music.

Not the flam of a single drum to time the steps. Our feet brush like loud whispers through the stubble of the field, or fall almost noiselessly on the turf of the meadow, or rustle through the leaves of the forest as our shoulders brush against the low-hanging boughs. The intense silence of the advancing line is more sublimely impressive than all the blare and crash of the noisy instruments of military music. We are marching into battle. The whole line is a living creature, with thought and feeling too profound for boisterous expression.

As the line moves forward a man occasionally lifts his head the least angle in the world and raises his eyes a trifle as they turn toward the center of the regiment. There, fluttering in the sunshine like a beautiful flower with wings and a soul, is what welds all the hearts in the regiment into one. No two men in the line could express their sentiment in the same phrase, but they all think the same thing. Any man who marches under that flag is worth dying for. The sun shines like a golden flame through a great rent in its blue field. That was a shell, gnashing its savage teeth as it tore through the galaxy of the stars. In the red stripes half a dozen stars of sunshine gleam. Those were Minie bullets that bit as they snarled through the silken folds. There are inscriptions, faint with many storms on the fluttering folds. The soldier knows the ragged letters by heart—"Iuka"; "Corinth"; "Jackson"; "Vicksburg." And to-morrow there will be a new name—fresh and clear. And a few names less on the regimental roster.

Every time Honor writes a new battle name in gold on the flag she blots the names of a few men off the regimental roll, in blood. That's the price of the battle inscriptions. That's what makes them so precious. The inscriptions are laid on in gold, underlaid and made indelible with blood. No wonder the Flag seems to be a thing of life. Every fold in it is aquiver with human hearts. When it is fluttering in the wind, it is throbbing. When it is unfurled in the rain, it weeps. The Flag—that is the Heart of the Regiment. And that it may never grow weak with the years and service, in every battle new hearts, young and brave and loyal, are transfused into the quivering veins of red and white; into the stars of gold on the field of blue. It is the living history of the regiment. It is the roster of the heroic dead, woven into the story of its many conflicts. It is memory and inspiration. It is the visible soul of a cause. So the men of the Union looked upon "Old Glory." So the men of the Confederacy gazed upon the "Stars and Bars" in the days of its hopes, when it flamed above fighting legions of the South.

I have seen it written that with the coming days of arms of precision and long range a general who would order his troops into action with a flag fluttering above the line to mark the location of every regiment would be court-martialed, charged with the murder of his men. Maybe so.

But I can't see how men could go into battle without the Flag to glance at now and again.

What reverence could a man have for a flag without a wound? How could you call a flag that wallowed its beautiful folds down in the dust all through a fight "a battle flag"? What is a flag for?

Why, when the bugle sounded the call for battle, quick as thought the color-sergeant loosened the lacing which bound the marching rain-proof case around the flag and the corporals of the color-guard snatched the covering off the National and the Regimental colors; the sergeants shook the beautiful standards out of their folds; the sunshine kissed them and the winds caressed them and tossed them in their arms—glad to see something as free as themselves released from the darkness. On the march the flag was cased against sun, rain and dust, that it might look brave as a bridegroom when it led the way to honor and victory. That was when we wanted to show our colors—when the enemy could see us.

"Here we are!" the Flag shouts to the skirmish-line, feeling its way through the dense woods hunting for us; "Here we are! This is My Regiment, right under my folds! Train your guns this way! You'll find us more easily than you can lose us!"

What Is a Flag For?

On every battle-flag might be inscribed a paraphrase of that splendid defiance of William Lloyd Garrison:

"I am in earnest—I will not retreat a single inch—and I WILL BE SEEN!"

That's what a Flag is for. How do you carry yours, Christian?

A man doesn't love anything or anybody very well unless he is ready to die for it.

Not necessarily to kill some one else, you understand. But to die yourself. To "present your body, a living sacrifice."

I suppose that is one thing that made the church so inexpressibly precious to the early Christians. So many people died for it. First, Christ, the only world conqueror in all history, the great Captain whose hand never curved around a sword-hilt, and who forbade his soldiers to slay or to smite. Then, generation after generation, the bravest soldiers the world ever saw, with peace in their hands and love in their hearts, met the armies of the nations, died for the truth and vanquished their persecutors, until the Cross gleamed in holy triumph above the circus of Nero and the Coliseum, and the Legions ceased to be. That is fighting love—the kind that conquers.

My regiment was one of the four which, with the Second Iowa battery, composed what is known as "The Eagle Brigade," from the fact that the Eighth Wisconsin Regiment of that brigade carried a young American eagle all through the war. "Old Abe" had the post of honor at the center of the regiment, his perch being constructed of the American shield, and he was carried by a sergeant between the two flags, the Stars and Stripes and the regimental standard of blue emblazoned in gold with the state coat of arms. All the brigade adored him, and "secured" chickens for him — he was fonder of chickens than the chaplain, and not half so particular about the cookery. To see him during a battle fly up into the air to the length of his long tether, hovering above the flags in the cloud of smoke, screaming like the bird which bore the thunderbolts of Jove, was to raise such a mighty shout from the brigade as would have blown Jericho off the map. Other regiments had dogs, bears, coons, goats. There was only one eagle in the army—"Old Abe."

He was an eaglet when the war broke out, and enlisted young, like many of the boys who loved him and fought beside him. He was captured on the Flambeau River, Wisconsin, in 1861, by a Chippewa Indian, "Chief Sky," who sold him for a bushel of corn. Subsequently a Mr. Mills paid five dollars for him, and presented him to "C" Company of the Eighth Wisconsin Regiment, known as the "Eau Claire Eagles." The soldiers at once adopted him as one of their standards, made him a member of the color-guard, named him in honor of the greatest of the presidents, and he never once disgraced his name. Through thirty-six battles he screamed his "Ha, ha," among the trumpets, smelling the battle afar off, fluttering among the thunder of the captains and the shouting. Never once did he flinch. He was wounded in the assault on

Vicksburg and in the battle of Corinth. At this battle it is said that a reward was offered by the Confederate General Price for the capture or killing of the eagle, "Pap" declaring that he would rather capture "Old Abe" than a whole brigade.

Sixteen Thousand Dollars from an Eagle

As he reenlisted at the close of his three years' service he went home on veteran furlough with his comrades, as he was entitled to do. When he said good-by to us his plumage was a beautiful dark brown from saber-curved beak to yellow shank. When he returned after sixty days, lo, he looked down from his shield in the majesty of a snow-white head and neck—more beautiful and regal than ever—the change that comes in the plumage of *Haliaetus leucocephalus*—that was his family name—at about three years of age. At the close of the war he was formally presented to his native state, Governor Lewis receiving him in the name of Wisconsin, from the hands of his comrades. During the winter of 1864, accompanied by a guard of honor, he attended the Sanitary Fair at Chicago, where the sale of his photographs, unautographed, netted the sum of sixteen thousand dollars for the fund for sick and disabled soldiers. He became a great traveler, being in attendance at many political conventions and soldiers' reunions. The sculptor, Leonard W. Folk, executed a model of him, which has been used in replica for a number of public monuments. He died on March 26, 1881, full of honors, though not of years, for he came of a family famous for longevity, some of his relatives living beyond the age of one hundred years. But his vitality was seriously impaired from the effects of smoke inhaled at a fire which occurred in his home, the state capitol in Madison, early in the year of his death. His body was prepared and mounted by a skilled taxidermist and occupied a prominent place in the military museum in the capitol until the building was destroyed by a second fire, February 24, 1904. "Old Abe" was a living standard, nobler than any effigy in bronze or gold ever borne above the legions of Rome or among the victorious eagles of Napoleon. It was fitting that his body should pass away in flames, even as the stormy years of his youth had been lived in the fierce joy that challenges death amid the fire and smoke of battle.

Dear "Old Abe"! I think of him every time I look at a quarter. His portrait makes it big as a dollar. I often wish all my creditors had belonged to the "Eagle Brigade." You see, patriotism not only makes a man's country seem greater; it makes her coinage appear more precious.

VII

COMRADES

It has been many changing moons since I attended a reunion of the Forty-seventh Regiment of Illinois Infantry. And I fear I may never attend another one until the Great Assembly. I would dearly love to. The "old boys" grow closer to my heart with every passing year. I was lonesome for a long time after the last reunion at which I foregathered with them. The years from eighteen to twenty-one are plastic impression plates of wax hardening into bronze with the years.

It is the Cause that makes Comrades. Not congeniality, nor personality. Comrades may be as antagonistic in personality as the sons of Jacob. They are church members, club members, Republicans, Democrats, Socialists, Insurgents, or any other human beings grouped together in one general class for high and earnest purposes. Brotherhood covers a multitude of sins—not wicked sins, you know, but disagreeable sins, which are worse because they are so much more numerous. Knowing who the dear Lord was, the society to which He was accustomed in heaven, its sweetness and purity, beauty and intelligence, I wonder many times how He could endure the disciples who clustered so closely around Him. I have sat in a boat on a warm day with Galilean fishermen on the Sea of Galilee. And they were no sweeter nor any cleaner two thousand years ago than they are to-day. I don't think our blessed Lord "liked" them any better than I did. But, then, He "loved" them. Which is quite different. You can't force yourself to "like" disagreeable people. But you can love them —dearly. For that is a command. And it's easy for a Christian to obey. It isn't for any one else; no. That's one of the tests of Christianity. I rather think it is the supreme test.

What Is a Comrade?

But in all organizations a "comrade" is a "comrade." That is the only definition. The dictionaries derive the word from the Spanish "camarada"; Italian, "camera"; English, "chamber"; French, "chambre"—"a military mess; those living in the same chamber or tent; an intimate association in occupation or friendship." But the meaning of a word, if it be a living word, isn't established by the dictionary. It grows, like a man. And how are you going to define a man? The dictionary says it is "an individual of the human race." "Specifically, a male adult of the human race." But that no more defines Ulysses S. Grant or Robert E. Lee than "a perennial plant which grows from the ground with a single permanent woody self-supporting trunk" defines a giant sequoia three hundred feet high, thirty feet in diameter, and seven thousand years old. You can't define "friend" in dictionary terms. And "comrade"—that isn't a name; that's a man. Tried by the acid test like pure gold, tried by the fire-test; by the wet fleece and the dry; by long marches; by hunger and thirst; by the long line of gleaming bayonets; by the thunder of the big guns; by the fierce reaping hooks of flame; by pain and wounds; by the fierce grip of battle; danger and death. That's what a Grand Army man or a Confederate Veteran means when he says "comrade." How are you going to put all that into a dictionary definition?

In the gray of early morning, in the quiet of noontide, or in the hour of the heaviest slumber, when the sky was the blackest velvet and the stars were whispering "sleep," the long roll broke into the silence like a storm of challenges, the men of your own company sprang into line, sent the cartridges home with swift dull-thumping strokes of the ramrod, and with sharp clicks of the hammers adjusted the caps and stood at attention, ready for anything and everything that might happen. You felt the light touch of the elbow that dressed the line. A quick glance between the men to note who stood next in line; a half-turn of the head to catch the face of your file-closer. You knew, then, that the man next you would be next you if bayonet lunge, screaming shell or singing Minie bullet found you;

that he would stop to pick you up if the line fell back, though the price of his stopping might be his own life; that he would spring to catch you if he saw you were going to fall, before you could call to him: that is "comradeship."

Yesterday you quarreled with him over some camp game. The day before, on camp guard, he dropped his musket "a-port" and barred your secret entrance through the lines, when detection meant disgrace and punishment for you. The day before that, when you were on "provost duty," you found him howling drunk and marched him to the guard-house, deaf to his piteous appeals to friendship. It wasn't many days ago you two fought in the Company street, as though there weren't plenty of chances to fight your common enemies. You never did like each other very well. He was a "moss-back Democrat" and you were a "black Republican"; he was a swearing, fighting, drinking scoffer, and you were a sober church member.

The long roll ceases, the colonel's "Forward, guide center!" preludes the explosive "March!"

and this man, as he steps out with you, gives your elbow an emphatic little touch that feels like a pat on the shoulder. Your mother wouldn't risk more for you that day than he will. She couldn't. And you know it. That's "camaraderie." Not a boisterous story or a rollicking song over a bottle of wine at night. But a sense of loyalty that lasts all day; that thrills in every nerve and throbs in every heart-beat. True "comradeship" claims all that one man has to give for another. The dear "Friend that sticketh closer than a brother," when He was giving to His beloved disciples a title dearer and truer than that of "brother," said, "I have called you friends"; "Greater love hath no man than this, that a man lay down his life for his friends." That's comradeship. That's greater than brotherhood.

"The brother," sadly said the Teacher, "shall betray the brother to death."

Cain and Abel were brothers. David and Jonathan were comrades.

"Blood is thicker than water," we all know.

Then there must be something thicker and warmer and redder than blood. A love truer than ties of kinship. A love that can do more than group a cluster of men into one family, or bind many families of men into one clan, or federate a score of clans into one nation. A love so pure and loyal, and so Christ-filled, that it will one day blend the whole world of men into one great throbbing heart of perfect friendship. Then will come the end of wars.

How Comradeship Was Tested

May 22, 1863, General Grant had made his march that opened the Vicksburg campaign, closing with the battles at Champion Hills, and the crossing of the Big Black. By the morning of the nineteenth a ribbon of blue, stronger than a web of steel, wound among the hills from river above to river below—Yazoo to Warrenton, and Vicksburg, like Jericho of old, was "straitly shut up." An assault upon the formidable works had been made on the nineteenth and had failed. But the soldiers were flushed with the succession of victories that had measured the march from Grand Gulf to the Big Black, and were not at all disheartened by one reverse. They "knew" they could take the city by assault—wanted another chance. And General Grant knew he could never keep an army in such a temper patiently in the ditches through the long operations of a siege, unless he first gave them their other chance and let them find out for themselves what they were up against and whom they were fighting. By the twenty-second all his troops were up and the second assault was ordered. You know more about it than I do, because you have read its many histories, and I was only in one little corner of it, very small, exceeding hot, and extremely dangerous, so that my personal observations, being much concerned with myself, were limited by distracting circumstances.

Anyhow, without much regard to my convenience, the assault was ordered at ten o'clock that beautiful May morning. Ten hours of the most terrific cannonading I ever heard; the assailing army storming the fortified position of an enemy almost its equal in numerical strength, when one man in a fort is considered the equivalent of seven assailants; Sherman, McClernand, McPherson, Mower, Quinby, Tuttle, Steele, A. J. Smith and Carr, war-dogs of mettle and valor. Hour after hour they charged the great bastioned forts, each time to be swept back with ranks thinned and scattered, but ready for another grapple. At half past three in the afternoon the brigade to which my regiment belonged—Mower's, then the third brigade of Tuttle's division, Fifteenth Army Corps (Sherman's)—was ordered, as a forlorn hope, to storm the bastion at Walnut Hills. We charged in column, and as we swept up the hill from the shelter of the ravine, we passed a little group of great generals watching us "go in"—Sherman, Tuttle and Mower, our corps, division and brigade commanders. Who wouldn't fight before such a "cloud of witnesses"? As we passed, Mower detached himself from the group and placed himself at the head of his own men. When we reached the crest of the hill we were met by a withering fire from the fort and stockade and breastworks that struck us in our faces like a whirlwind of flame and iron. We fought through it, close to the fort, when we were finally repelled. Then there happened to me that to which the rest of the day's fighting seemed only preliminary.

Bringing Back a Lieutenant

As we fell slowly back, I saw our second lieutenant, Christopher Gilbert, stagger and fall crookedly forward. I thought he was killed, but as I looked for a moment I noted him trying to rise. It wouldn't do to leave him there,—that was certain death. Robley D. Stout, one of my company, and I ran to him, and lifting him to his feet, drew his arms over our shoulders, and brought him back to the retreating-line. He was shot through the leg with a grape-shot, and unable to help himself more than to cling to our shoulders. I wished at the time that he were as big as a bale of hay, for his body made a sort of shield for the two youths who were carrying him away from the missles that still pursued him spitefully as though they were bent on finishing the work they had begun.

He recovered after a tedious time in hospital, and when he could return to duty the additional bar he won at Vicksburg graced his shoulder-strap, and he was our first lieutenant. There were two Gilberts in the company, Chris and Charley, brothers, good boys and good soldiers. I met my lieutenant a few times after the war. Then our lives drifted apart. I became a minister and was pastor of Temple Baptist Church in Los Angeles, California.

And one day my lieutenant came before me, not to give orders, but to take them. He was a prisoner, and his fair captor stood beside him. She had done what Pemberton's sharpshooters in Vicksburg could not do. Love had won my lieutenant. I ordered him to accept the terms of the bride, to "love her, comfort her, cherish her, honor and keep her, till death them did part." And he obeyed willingly.

COMRADES

After the service he said:

"Bob, do you recall the hot afternoon on the slopes before the bastion at Vicksburg?"

"I was just thinking of it, Lieutenant. And I was wondering if now you might ever blame me for helping to drag you out of the range of Pemberton's sharpshooters?"

"Indeed, no," he said, "I never will. I've often wondered why the dear Lord sent you back after me. But this is the 'Why.' "

And I guess it is, for they have entered into the supreme comradeship, "wherefore they are no more twain, but one flesh."

VIII

THE TESTING AT THE BROOK

"THAT's a fine army," said Gideon, a general appointed from civil life—what our West Pointers call "a mustang," a good horse with no pedigree, a good soldier without a West Point diploma,—"that's a fine army," looking at his first command, and the largest he had ever seen; "thirty-two thousand able-bodied men. I can whip Midian off the map with these heroes."

But God, who had seen many armies, said softly to Himself, not to hurt the general's feelings, "Not with that crowd you can't." Then He commanded: "Send home all your cowards."

And the general, who didn't believe there was one in his army, forgetting that he himself "feared his father's household and the men of the city" when he half-disobeyed an order, called on those who were "fearful and afraid" to strike for home and mother when they were ready. To his amazement, two-thirds of his corps, twenty-two thousand men, catching sight of Midian encamped in the valley, made an early start from Mount Gilead before a bowstring was tightened, and got home before the war began. And Gideon reviewed his remaining ten thousand.

"Well," he said, "this is as big an army as Barak had when he destroyed the hosts of Sisera. Much may be accomplished with ten thousand selected men."

But God said, "A phalanx is better than a mob. Try them out at the ford of the brook, and keep all who really want to fight."

And of ten thousand soldiers there endured the final test three hundred fighting men,—men so hungry for a fight and so eager to find an enemy they forgot they were thirsty when, with parched lips, lolling tongues and panting breath, they went splashing through a desert brook on their way to battle.

Sweating Down to Fighting Weight

So from that day to this every army has had to be sweated down to its fighting weight by similar, although slower, processes. There's a lot of useless material about an army; about as useless as noise is to a wagon. And yet wherever the wagon goes the noise goes. Look at the stuff a church gathers about itself at the end of a six weeks' revival. Wait until the revival is six months old, and the church has been fighting sin in all its subtle forms every day in all that time. Then call the roll at the prayer-meeting. And yet most heartily do I believe in revivals. But the great net brings to shore lots of fish that are good for nothing but to cast away. And of all things that are a revolting stench on dry land, a worthless fish is a little the loudest and worst.

Of the army of Israel's deliverance only three hundred were "Gideon's men." In our modern wars much the same thing is approximately true. There isn't much new found outside of the old Bible, is there? A thousand men enlisted in a regiment in 1861. In 1862 the regiment counted itself strong if it carried three hundred bayonets into battle. These three hundred constituted its fighting strength. The line on dress parade no more represents the regiment than the big well-dressed congregation Sunday morning represents the church. All the skulkers appear on dress parade, usually in the smartest uniforms—they do nothing to soil them—and in the front ranks. The fighting men do not show at their best on parade. Some of them were killed in the last fight. Others are in hospital, nursing their wounds in the ambitious hope that they may rejoin the regiment before the next battle. A prayer-meeting is never so showy as the Sunday morning congregation.

Considering the fact that the world has been at war ever since there were three men in it, comparatively very few military organizations have left a record for courage of the highest type,—what Napoleon Bonaparte called "two o'clock in the morning courage": courage that is just as trustworthy, clear and sane in a sudden emergency as with the average time of ample preparation, knowing at once what to do and fearlessly ready to do it. It was the blare of the trumpets and the glare of the torches at midnight that defeated Midian, before ever a blow had been struck with the sword. Of the famous troops, one thinks at once of Gideon's three hundred; David's mighty men, although theirs was an example of individual prowess, rather than the achievements of a band; Leonidas and his three hundred Spartans at Thermopylæ; Napoleon's "Old Guard"; Cromwell's "Ironsides," of whom he wrote, "truly they were never beaten"; the six hundred at Balaklava; Frederick's "Grenadiers"; and, of course, the regiment that went from your town in the sixties. I should have chronicled that one first, but I couldn't think of the name. The fighting "Onety-Onest," wasn't it? It was "The Fighting" something, I know. That's what it called itself. I belonged to that regiment myself.

Were there no cowards in any of these famous organizations? There may have been at first, but they were sifted out. But of those hard fighters who were left, were there none who were at times a little bit frightened? Was there ever a soldier who was never "afraid"? After the battle of Kunersdorf, Frederick the Great was as nearly scared to death as any man could be and not die. Napoleon, who died an exiled prisoner, should have fallen at Waterloo. Maybe he couldn't. Carlyle tells of Ney, whom the emperor called "the bravest of the brave," raging through that fearful carnival of death crying, "Is there no bullet for me?" Cæsar, Hannibal, Napoleon, Wellington, Grant, Lee, Sheridan, Jackson,—did these great captains never feel the sense of fear? Their critics will answer "yes," their admirers, "no." If I have the casting vote, I will have to vote with the "ayes."

And why? Well, because they were men before they were soldiers. They were human beings. And if I may judge from my own limited and narrow experience, ranging through one generation and all around the world, I have never met more than a half dozen men who declared they had never felt the sense of fear. And none of these was a soldier, and all of them were liars.

And another reason I have for thinking these great captains knew what it was to be afraid, is that they were splendid soldiers and brave men. And no man reaches the highest point of courage who has not overcome fear. Fear—it is a part of our humanity. In the truest story of the race that was ever written, fear is named before love or hate: "I heard thy voice in the garden, and I was afraid." So the first man that ever lived was tormented by fear. And Abraham was wounded by it. And Jacob. And Moses. And David. And His mightiest soldier, Joab. And Peter, the bravest of the Twelve. And Paul, the great apostle. I tell you, man, if you have never known fear, your courage has never been tested. The bravest men are converted cowards.

The Sifting Worse Than Fighting

What is a coward, then? The sort of man who is disgraced before the regiment because he has dishonored the colors,—whose military buttons are cut off, whose head is shaved, and who is drummed out of service down the length of the line on parade to the *Rogue's March?* What makes him a coward?

He was all right when he enlisted. He knew as much, which is to say, as little, about war as the rest of us. He knew that a soldier was mighty liable to get shot. He counted the chances. He was a patriotic citizen. He loved his country well enough to offer and to risk his life for her. What made him a coward?

Well, the sifting process is something terribly drastic. It's worse than fighting. The men who failed at the brook test could have gone through the battle with the Midianites all right could they have crossed the brook in the right spirit. The awful quiet before the battle; the muffled hum of preparation; nothing to do but form and wait, with plenty of time to think about it. And the environment does not suggest pleasant lines of contemplation. You note how small your own regiment is—two hundred men. But you picture the Confederate Eighth Georgia, right in your front, with nine hundred fighters. Really it isn't so large as your own, but you don't know that. As the battle draws nearer and closes around you, a score of things happen to "scare" a man, even though he may be a brave soldier. And they scare the coward a great deal more.

As you lie on the ground to hide the position of the regiment from the enemy and to keep underneath the searching shell-fire and the skirmish shots that get past your skirmishers, a man is talking to you, with his face turned toward your own, a foot away. You are listening to him with interest, because he is asking you about something that happened in your own town, in the Lincoln campaign. As you start to answer him, something fearful blots out his face with a smear of blood, and he is a shuddering thing without voice or breath or soul, huddled there at your side. A shell has burst above your company and a piece of it struck that man in the face like an angry specter that resented his question.

Before you get over this, as you look along the line for encouragement in the faces of men braver than yourself, your eye catches the glance of a soldier four or five files away. A smile plays on his lips in answer to your glance, which he rightly interprets. Then suddenly his face whitens like death. He lifts his head a little; his open mouth gasps for air—once—twice. Then he lays his face back on the grass as quietly as though he were going to sleep. A bullet hissing along through the grass like a lead serpent had just found his heart.

And This Is War

Another shell bursts over you with a sudden shriek and a cloud of stinging smoke that burns your eyes, and half a dozen ragged fragments hurtle through the blue dusk. One of them snaps like a mad dog at the foot of a comrade, tears and shatters the tiny bones of ankle and foot. The man screams with agony. As they carry him away, he troubles the air with his cries, for he knows he is a cripple for life.

The regiment rises to its feet and begins at the command " 'Ten-shun! Commence firing!" The man next you who has closed up that first hideous interval is vigorously ramming a cartridge down his rifle. He says to you, "What shall I do? This cartridge is jammed!" "Spat!" a spiteful Minie ball interrupts him, as it crushes the elbow of the lifted arm with a sound so cruel that you flinch with the other man's pain. But he—he twists his face into a grimace to hide his hurt and answers his own question, "I won't do anything with it!" as he walks back to the rear, a one-armed man forever.

Sometimes a tragedy has a ghastly sense of wonderment that is near to grim humor. In the assault on Vicksburg I saw a comrade stoop twice in vain efforts to pick up his musket, knocked from his grasp, before I could call to him that his hand was gone. A bullet had cut away every finger on his right hand, and all that he felt was a painless sense of numbness. My face must have showed what I thought, for an older soldier, laughing as he capped his musket, said:

"That's all right, Bobbie; you're liable to get killed any minute and never know a thing about it!"

That's what I thought. But his confirmation wasn't half so reassuring as it sounded.

Then, again, as the regiment stands in line waiting for the "Forward" that will send it like a whirlwind upon the battery in its front, a great solid shot with a devilish shriek, wasting its mighty force on a life that a tiny rifle bullet could destroy as completely, smites a man in the chest, and hurls him twenty feet out of the line, tearing him to pieces like a wild beast.

Now, if after all these things you still want to fight, if you shout loud and long and exultantly as you spring forward to follow the flag when it advances, then you have got across the brook with only one or two refreshing laps at the cool water that bubbled in alluring crystals around your knees. Now you can be trusted with trumpet and pitcher and sword. Now, "faint, yet pursuing," you will hang on the trail of your beaten enemy like a hound on the trail of the wolf.

But the hardest fight took place and the great victory was won when you fought with yourself as you splashed through the brook with your head in the air.

IX

THE COWARD

Why does the coward go to war?

It is the most dangerous occupation in the world. It offers the greatest discouragements and the smallest rewards for cowardice. It most emphatically professes for timidity only unmeasured contempt. It is the calling for which the coward is most unfitted by temperament and inclination.

Why, then, does the coward even start to war?

For certainly he does start, in every war that is declared. He is found in every army. He goes to war voluntarily, many times eagerly, for the cowardly temperament is volatile. A rabbit is sprightlier than a bulldog. The coward may start to war with the valor born of ignorance. When I enlisted, I had but one well-defined fear. I was afraid the war would be over before I got into a battle. Every time I got hold of a newspaper or news reached the camp by courier, my heart sank with the disloyal dread that that old Grant—all generals are "old" to the soldier—had utterly crushed the enemy with one terrible blow, and I would have to go home without one battle story. It was terrible. However, it didn't happen. Though many a time afterward I wished that it had. I got into my battle. After that a second fear displaced the first. I was afraid the war would be ended before I got into another. And again my fear was an illusion. The war kept on until I got into a score of fights. And then, seeing perhaps that *I* was never going to quit first, the hosts of the Confederacy agreed to stop if I would. At least that is the way it appeared to me. And it seemed to be an honorable termination of the prolonged and obstinate struggle. Up to 1865 I had killed as many of them as they had of me, so that honor was satisfied. And you couldn't tell what might happen. At the next grim roll-call of artillery and musketry, they might get the delegates and have a majority of one over me. Not much of a majority, but it would be as good as unanimous.

The Business of Fighting

There is more "thrill" in the first battle than in any of the subsequent ones. You may go through harder battles than the first; longer, fiercer, more savagely contested, bloodier in every way. But no soldier will ever forget item or incident in his baptism of fire and blood. He fought like a patriot. He never forgot the high and holy cause in which he was a soldier. He looked at the flag with his soul in his eyes. He cared no more for his own life than he did for the grass under foot. Older and braver soldiers than himself reproved his recklessness. He was daring without cause. He stood up like a man, aimed deliberately into the smoke that concealed the hostile line, and hit a tree-top. A man isn't afraid in his first battle. He is excited; thrilled, his nerves are on a tension like harp-strings; his senses are abnormally alert; he sees everything; hears everything.

In all the others he fights like a soldier. He takes sensible care of himself. He may fire many times without seeing a man. But every time he shoots into a place where he knows there are men. He fights a little better every time he goes into a new battle. A certain commonplaceness of war infects him. He's where his business calls him. He chatters about all sorts of things, because his nerves are tuned up to concert pitch; but he isn't nervous. He discusses with his comrades the merits of the campaign of which that battle is the key-note. They dispute about the weight of the guns that are shielding them; they distinguish between "rifle" and "smooth bore." They listen to a report—sharp and clear—that splits the battle clamor like a new voice and say, "That's a Rodman"; and they hear a great boom, loud and heavy, and say, "That's a bronze Napoleon." Thus they introduce the recruits to the machines that are trying to kill them. They draw cuts with blades of grass to see who shall take the canteens and hunt for water. And when the unlucky one returns and some foolish one asks, "Where did you get the water, Bill?" there is a roar of laughter when Bill replies, "Don't ask me till you've had your drink." They imitate the whining of a bullet that comes unpleasantly close, and echo the shriek and howl of the shell. Some of the men are such artists in this mimicry that their efforts are encored. They reply to the shell that comes along with its "whoo-whoo" with innocent answers—"Who? Cloyd Bryner? That's him—four files down the line." "Who, me? I'm not here. I'm back in Illinois. Ain't never been out of the state." They chaff one another's personal peculiarities to the verge of a quarrel. They act like the bleachers at a baseball game, or football players between the halves. That's their business—fighting.

If you can get the coward safely into that, he'll stay and he'll fight. As a rule he fails in the preliminaries. But sometimes he gets so nearly across the brook that he has only one foot in the water—and then he lies down for a drink.

I remember a coward whom I knew in the army.

A good coward. In all other respects, a good soldier. A pleasant-looking man, with a weak chin, hidden by his long beard. Blue eyes, kindly as a woman's! A manly voice; an intelligent mind. A cheery comrade; rather quiet. Never shirked a duty in camp or on the march. Neat in his dress; excellent in drill. Gun and accouterments always bright and clean. In scant-ration times, always ready to divide what was left in his haversack or canteen, taking the smaller portion himself. Vigilant on camp-guard, though I soon observed that he was never detailed for picket duty, where a man may have to stand vedette—away out by himself, with his own responsibilities—a very lonely post of the highest importance.

This man was a coward.

He knew it. He was ashamed of it. He tried to overcome his cowardice. The regiment never went into battle that he didn't start in with his company. If his number brought him into the front rank, there he stood. He rammed down his cartridge with a look of resolution on that uncertain mouth, and he "fixed bayonets" with the air of a man who is going to reach somebody with it, in spite of the modern military axiom that "bayonets never cross." He lifted the hammer twice or thrice to be certain that the cap was good and fast on the nipple. He tightened his belt a hole or two, as a man who knows there is going to be hot work and no dinner-hour. He shook his canteen at his ear to be sure there was a good supply in case he was wounded. He made all the preparations of an experienced, "first-class fighting man" who intended to volunteer when a forlorn hope was called for some desperate duty, on which only picked men would be taken.

What Happened Under Fire

And his comrades stood by him and helped him, for his reputation was known, his weakness and his good points. A sergeant fixed one eye exclusively on him. His nearest comrade touched elbows with a little ejaculation to "play the man." The captain paused behind him as he walked down the line and whispered to him. The lieutenant caught his eye and nodded encouragement. Unconsciously we all seemed to be leaning a little closer to him. Then the order translated the bugle with a shout, the flag fluttered and the line moved forward; a rain of shots told that our skirmishers had found them, and just as we were ready to dash forward like dogs of war the man nearest the coward stopped, choked, coughed up a stream of blood and fell sidewise.

And the coward ran away.

Broke from his file-closer who tried to stop him; tore loose from the corporal who clutched his arm; threw down his gun; dodged the sergeant who lunged fiercely at him with his bayonet; out-stepped the lieutenant who ran after him; ignored the wrathful shout and threatening revolver of the colonel, and was safely gone. That was as far as ever we could see him. Back to the rear he raced. Past the supporting lines; back into the ruck and rabble of other cowards and the demoralized horde of camp-followers that make the rear of the fighting line a pandemonium of fear and misrule and confusion, despite the good soldiers held there on duty. He ran away.

Sometimes shame kept him away from the regiment for a day or two, or even three. But he always came back with a wild excuse for his disappearance which we all knew, himself included, was a foolish lie, and resumed his duties.

In the first instance he suffered for it. The regiment resented it. His company felt disgraced. But insensibly our attitude toward him changed. Cowardice is one of the most serious offenses in the army. It is punishable by extreme measures—even death. I have seen men "drummed out" of the service for it. But no charges were ever brought against this man. He was never punished. And being a young soldier when I joined the regiment, I used to wonder why. And often I wonder about him in these quiet days when a saluting cannon on some day of parade sets my heart beating for a moment with quickened throbs as I half listen for the exploding shell to follow. No man who ever had a loaded gun fired right at him ever again hears rifle or cannon-shot with the same indifference that the civilian feels and shows. It is exactly the difference in the looks and feeling of the man who loves to read and the man who can't read a word, as they walk past the shelves of a library.

Was He a Failure?

But whereas in the fierce old days I wondered why the colonel didn't court-martial the coward for running away, I now wonder if the man was a coward, after all!

For the cowards all ran away before the battle, when they didn't have to run. They went back from Mount Gilead as soon as they saw the enemy. They stayed away. They played sick the day before. They fell out of the marching ranks when we began to double quick. They stopped at the fence when the regiment suddenly deployed into line to tie up a shoe that was already so knotted they couldn't untie it. They got details in the hospitals in St. Louis and Cincinnati and other northern cities months before. There were scores of ways of keeping out of a battle without actually suffering the charge of cowardice. And some there were who ran away on the way in, who got so far across the brook they could hear the distant batteries and the nearer skirmishers.

But this man went in every time. With what beating of heart, and straining of nerves, shortness of breath, and strenuous calling up of all the reserves of resolution and will-power, God knew, and the colonel half guessed. A braver man, up to that point, than any of the rest of us. He started in, and he would have stayed through but for that awful smear and sickening smell of hot blood. If we could have held him past that, either he would have won his chevrons or died of heartsickness. Somehow I think if the coward, when he went to enlist, could have got a message through to the dear Lord, and had waited for the answer, and could have understood it, he would have been told that "they also serve who only stand and wait." God never intended that man should kill anybody. I have known other men since those days, calm-natured, fearless, who can not abide the sight of a tiny splotch of blood. They faint, as they look, as though the surgeon's steel that drew the crimson drops had pierced the patient's heart.

The coward served through the war, and when the regiment marched home to welcome and honors, I think one of the bravest men that went with them was the coward. I know he was beaten in every fight he went into, but he went in. And he fought. And such fighting! Much we knew about it, we laughing, shouting, devil-may-care schoolboys playing with firearms!

What is a coward, anyhow? Cravens, and dastards, and poltroons, we know at sight. But who are the cowards? And how do we distinguish them from the heroes? How does God tell?

X

THE SCHOOL OF THE SOLDIER

In the old Hardee tactics the "School of the Soldier" is the title of that chapter which pertains to the instruction of the individual soldier in the things which he can do by himself—the manual of arms, for example, which he can study and perform in solitude as well as in the company of the regiment, with the excepting of "stacking arms," which no soldier can do with one musket. And the facings—right and left and about, and things of that sort. After he has passed his finals in this school there comes the "School of the Company," and of the battalion and other larger movements and evolutions. He must have his entire company to assist him to form fours or march by the right or left flank as the case may be with or without doubling. He can not deploy as a skirmisher without at least a platoon to cooperate, nor can he "rally by fours" without a minimum support of three other warriors, nor "rally by squad" without comrades to the extent of a group. And when he advances in the broader knowledge of battalion instruction he learns how impossible it is for one man to form a square, with officers in the center.

There are some things, you see, that a man can do by himself. He can learn to handle his rifle; stand sentry; go on certain phases of fatigue duty, chopping wood or policing the color-line—oh, lots of things a man can do by himself. Guard his heart and keep his lips; brush his teeth and control his thoughts. All these things pertain to the man. In the army and in the church and in business, certain things belong to the "school of the soldier." But in the large things of life, you have to work with human cooperation. All sorts of humans, too. Some of these things—maybe, I am not sure of that—pertain exclusively to the army and military education. Most of them, perhaps all of them, are useful in every calling. Some of these things we are taught by our officers.

THE SCHOOL OF THE SOLDIER

Not Taught by the Officers

The president of our division, our general, taught us in great masses and large movements which we did not understand, but which we knew he did, and that sufficed us. As for personal instruction, almost any good soldier in the ranks could have corrected the general in his manual of arms. He taught the "larger good."

The colonel came a little closer to us, in the regimental drill. We learned the reasons for his movements and responded to his commands with springing alacrity because his eye was upon every man in the line. The captain was a rigid catechist, knowing each one of us, his catechumens, and teaching us with the keenness of a martinet the things that pertained to our personal military salvation. But closest of all, the old-fashioned sergeant stood only a ramrod length away from the individual, with an eye for dust and spots like a hawk for prey. His was the "bark" with a bite behind it that straightened the slouching shoulders and brought the wandering little finger back to the seam of the pantaloons. His the mechanic's glance which "dressed" the company line till it answered the geometrical insistence of the shortest distance between two points. His the sarcastic taunt which adjusted the awkward feet until the heels clicked together and the toes pointed at the required angle. The sergeant was the man who "licked into shape" the shuffling recruit and made a soldier of the cub of the awkward squad.

In the church the pastor is the general, the superintendent is the colonel and the Sunday-school teacher is the sergeant.

The colonel taught us things the general had no time for. The captain taught us some things the colonel couldn't. The sergeant taught us nearly everything the higher teachers left out.

Then there was the best teacher of them all—the pedagogic martinet without warrant or commission, chevrons or shoulder-straps; the old-fashioned birch-rod teacher who taught by main strength, precept and example—the private gentleman of the line. He never assumed to give instruction to the company or battalion. But as an unassigned instructor in the "school of the soldier," the private taught his fellows their lessons without opening the book, and graduated every pupil *cum laude.* There's a heap of things you learn in the army—and in civil life—that are not in the book, and nobody can teach them so well as the other soldier.

Send the boy to kindergarten at four years. At six he goes into the primary. Ten finds him in the intermediate. Fourteen sees him in the grammar-school. At eighteen he is graduated from the high-school, and at twenty-two he comes out of college and passes a few years in the university. Then a couple of years in Europe, and at last his education is complete as far as the schools are concerned. So much for the books. But, oh, the things he learned from the other boys! The unwritten things—these form the important part of his education.

It was the private soldier who taught me not to step on the heels of my file-closer. He also taught me how to make a feather bed of two oak rails. How to grind coffee in a tin cup with the shank of a bayonet. How to boil roasting-ears in their own husks in the ashes. How to drink boiling coffee without blistering my throat. How to conceal my person behind a sapling not half so thick as my body. How to fill my canteen from a warm pond and let the water cool in the sun on a hot day. How to march eighteen or twenty miles over rough roads day after day without getting an ache in my feet. How to make one day's rations last three days without going hungry. How to get a refreshing drink of water without swallowing a drop. How to lift a nervous hen from the bosom of her family without any outcry from herself or relatives. How to fool the sergeant on roll-call—once. That trick was like a limited ticket, good "for this day and train only." How to "explain things" to the captain. How to launder one's linen, which was woven of the coarsest flannel, in cold water. How to make one's self clean when it was muddy, and how to look fresh when it was dusty. How to divide the last pint of water in your canteen so as to get a drink and a sponge bath and have enough left for coffee.

How to make two months' pay—twenty-six dollars—last till next pay-day, two or three months away, after you had sent half of it home and spent half the remainder. How to keep awake on picket all night when your dry eyes ached and burned for sleep. How to sleep like a tired working man under the guns of a battery shelling the enemy's lines. How to light a fire in the woods with wet twigs in a pelting rain and a fretful wind with your last match.

When the general, colonel, captain and sergeant have done their best for you, you turn to an old private soldier in your own company and say:

"I have been graduated in the school of the soldier and have a diploma signed by four able instructors."

The Orderly and the A. D. C.

And the private says:

"Very good, my son. You have been a diligent pupil. Now come to my high school and I'll teach you something worth knowing."

Oh, you do learn something from what you read! "Reading maketh a full man." And you learn a little more from what you see and hear. But you take honors in the things you do. It's a mighty good thing to "know these things," but "blessed are you if you do them." For then you will know how to do them right, and you'll never forget them.

Once upon a day there came to our regiment, on the march, a staff officer splashing along the rain-soaked road, trailed by an orderly, who looked twice as important and rode much better than the officer. The aide-de-camp checked his galloping steed in front of Colonel Cromwell, showering the regimental staff with yellow mud and water as he saluted. The orderly halted quite as abruptly, but did it much better, splashing no one. Because he had been trained in the school of equality, which has been the training-school for humble folk and plain people in all ages. He also used to splash the infantry colonel when he rode up in mad haste.

That is, once he did.

And he was also wont to halt before a platoon of cavalry with the effect of a mud volcano in active eruption.

That is, he was wont to, once.

And also, he was once accustomed to ride splashfully along the line of a marching regiment of infantry, impartially distributing emulsions of mud and water as the churning hoofs of his charger might direct.

That is, once upon a time he did. The next time he didn't.

That is one of the excellencies of the educational processes of the public schools. And the army is a public finishing school. The officer, like the son of a millionaire in a private school for young gentlemen, can take many liberties and do many impertinent things. The private soldier, being promptly admonished by his comrades on his first offense, which is considered as heinous as his last, usually makes it his last. He may offend in other ways, for there are a thousand ways of being mean. But rarely, if he is a good soldier, which means if he has common sense to begin with, does he repeat his first crime.

Where the Training of Adversity Wins Out

Wherefore on this occasion, the orderly, having aforetime been bombarded by infantrymen of his own rank, and slammed by cavalrymen of his own grade, remembering that it was just as easy and far more acceptable to the audience to splash the bank of the road as the faces of his comrades, did so. And was loftily and sternly insensible to the sarcastic eulogiums upon his skill and tact, uttered in loud tones by the soldiers web-footing along the middle of the road.

Thus may we see how sweet are the uses of adversity and the lessons of poverty. The man who learns to bear the yoke in his youth doesn't mind the callosities in his age. Nobody so wisely instructs us in the practical ways of life as our equals. I learned a hundred things, more or less useful, from my comrades of the rank and file that my colonel never taught me. Indeed, some of these accomplishments he sternly and faithfully admonished me to forget.

Meanwhile the untaught aide-de-camp has saluted and galloped away in showers of mud, followed, at safe and dry distance, by the orderly, who rode well and wore the bearing of a division commander, piling it on a little bit high, perhaps, because he was aware that his immediate superior rode very badly, his awkwardness being emphasized by the floundering of a gaitless horse. The orderly, who was a chambermaid in a livery stable when he enlisted, had gone to the corral when the new horses came in and picked out his own from a couple of hundred. But the aide-de-camp, who was a bookkeeper in a shoestore when the war broke out, had sent his colored servant to select a horse for him. The negro man, before that he was a freedman, had been a plowboy on a Mississippi plantation, and his idea of a good horse was a brute with bunchy knees, big hoofs and tremendous quarters. And that was the style the aide-de-camp got. Hence the contrast between the riding of the A. D. C. and the orderly.

Oh, you go to school to learn to read and you go to college to "increase learning"; you pore over the endless output of books to enrich your mind, and you burn the midnight oil in the "much study that is a weariness of the flesh." And all this time "Wisdom"—who is quite a different thing from books and lectures—"Wisdom crieth aloud in the street; she uttereth her voice in the broad places; she crieth in the chief place of concourse; at the entrance of the gates"; in the ranks of the private soldiers; in the crowds of the common people—where there is dust, and care, and toil, and poverty; pain and heartache; fighting and dying. That's where you find "Wisdom."

XI

GOOD FIGHTING ON POOR FOOD

Two fierce October days we fought with Price and Van Dorn at Corinth, Mississippi—the third and fourth. Twenty-eight thousand Confederates hurled themselves against the forts—Robinette, Williams, Richardson, and a fourth, Powell, near the Corinth Seminary. These were all new forts, unknown to the Confederate generals. Fort Williams was a very strong work, defended by big thirty-pound Parrotts—a type of our best guns at that day. I have heard of surprises in war. But I think the completest I ever saw was one sprung by this bastion of big guns upon a little Confederate field battery at the beginning of Corinth fight. The enemy knew well the location of the town, and before daylight they ran that little park of six-pounders up under the guns of Battery Williams and began shelling the town, greatly to the discomfiture of a few thousand non-combatants who dwelt therein, for it was a great depot of supplies. The gunners enjoyed the little fight they were having all by themselves, while the big fort, blanketed in the darkness, emulated "Brer Rabbit" and lay low. But when daylight raised the curtain, those thirty-pound Parrotts, as though amazed and indignant at this nest of popguns playing war on the very door-steps of their fort, tore themselves loose like rifled thunder-storms at musket range. The little Confederate battery? Oh—that? One evening of a dismal November day a kind-hearted citizen of Brooklyn observed a little boy, weeping bitterly, standing at the foot of a dark stairway that led to some mysterious region above.

"What is the matter, little man?" asked the citizen kindly.

"Pap's gone up-stairs to lick the editor," sobbed the boy.

"Well," said the philanthropist, "hasn't he come down again?"

The boy sobbed afresh. He said:

"P-pieces of him have!"

It seems to me, as I recall the incident at this distance, that fragments of those little cannon came down during that afternoon. However, I was much occupied with things nearer me that day and was not looking for the descent of terrestrial meteors.

A Baptism of Fire—and of Tears

Twenty-eight thousand Confederates dashed themselves against our line of defense those two savage days like waves of the sea. My own regiment lay in the ditch of Battery Robinette, which bore the brunt of the final attack. Curtains of infantry connected the forts. For a wall of sand is as good to stop the sea as a sea-wall of granite. Twenty thousand boys in blue there were under Rosecrans, fresh from fighting the same foes at Iuka, where our major, Cromwell, had been taken prisoner. The fighting on the third at Corinth punished the Federals severely. At half past nine o'clock on the morning of the fourth, Price's column, formed en masse, came charging along the

Bolivar road like a human torrent. It moved in phalanx shape through a storm of iron and lead from batteries and infantry, and drove through all opposition, the men bowing their faces, but pushing on, as men crowd their way against a driving storm. As it came within rifle-range the phalanx divided into two columns covering the front of the forts. It captured Fort Richardson and General Rosecrans' headquarters, in front of which seven dead Confederates were found after the battle. It seemed that nothing could stop that onrush of determined men. But in the score of minutes that so often decides a battle, the Fifty-sixth Illinois recaptured Battery Richardson, the heavy assaulting column was thrown into confusion, and the splendid charge was turned into a swift retreat. The whole affair lasted half an hour.

Meanwhile Van Dorn's column, which should have cooperated simultaneously with that under Price, but was delayed by the natural obstacles of broken ground, tangled swamps and densely-wooded thickets, came charging in on the Chewalla road. Texans and Mississippians these fighters were. I was greatly disturbed to perceive they were headed straight for our position—Forts Williams and Robinette; but then I thought of those fearful Parrott thirty-pounders and the terrible guns of our own Robinette trained point-blank on that charging whirlwind. Colonel Rogers himself led his Texans, densely formed, in a close charging line massed four deep, the Mississippians keeping pace with them. The infantrymen sprang to their feet. Volley after volley of musketry helped the big guns tear the assaulting lines to pieces. But they kept on. They struck the infantry supports as a great combing wave strikes a reef. They beat us down with their muskets and thrust us away with bayonet lunges. Colonel Rogers leaped the ditch at the head of his men and was killed on the slope of the parapet. We saw the soldiers in gray swarming into the embrasures, fighting with the gunners who met them hand-to-hand with muskets and sponge staffs. The Ohio brigade of Stanley's division, firing withering volleys, came to the rescue of the forts and their supports, and Confederate reinforcements hurried into the maelstrom of fire and steel. Our colonel, Thrush, was killed, shot through the heart. Step by step we crowded them back until they shared the fate of the other column and turned in retreat. The battle was over. Battalions of gray and blue stretched themselves along the dusty roads toward the Hatchie River, in mad retreat and hot pursuit. Of the forty-eight thousand troops engaged, seven thousand two hundred were killed and wounded, showing how continuous had been the fighting. There had been no idling precious time away in the great industry of Christian nations—killing one another. Of the casualties four thousand eight hundred were Confederates, two thousand four hundred Federals. I know that none of the wounded, and I don't think one man of the killed on either side, changed his opinions because the other man had fired first or more accurately than himself. That shows how much of an argument war is.

That night I was detailed on duty with the parties that go over the field, looking for the wounded and the dead, succoring the living, burying the dead. The savage day had been a baptism of fire. The night was a baptism of tears. The day had been the terrible inspiration of battle. The night was the meditation of sorrow. On the battle-field Death was the grisly King of Terrors, wearing the black plumes of a mighty conqueror, naked and splendid and bloody in his brutality. Fighting under his crimson standard the gentlest soldier was shouting, "Kill! kill!" Here, in the starshine that sifted sorrowfully down through the pines on the white faces and mangled figures, he was terrible in his silent reproaches—"Why have you men called me out and set me on to do these things?"

The Acorn's Silent Message

We found a dead Confederate lying on his back, his outspread fingers stretched across the stock of the rifle lying at his side. He was one of Rogers' Texans. Fifty-seven of them we had found lying in the ditch of Battery Robinette. I covered his face with the slouch hat still on his head and took off the haversack slung to his neck that it might not swing as we carried him to his sleeping-chamber, so cool and quiet and dark after the savage tumult and dust and smoke of that day of horror.

"Empty, isn't it?" asked the soldier working with me. I put my hand in it and drew forth a handful of roasted acorns. I showed them to my comrade. "That's all," I said.

"And he's been fighting like a tiger for two days on that hog's forage," he commented. We gazed at the face of the dead soldier with new feelings. By and by my comrade said:

"I hate this war and the thing that caused it. I was taught to hate slavery before I was taught to hate sin. I love the Union as I love my mother—better. I think this is the wickedest war that was ever waged in the world. But this"—and he took some of the acorns from my hand—"this is what I call patriotism."

"Comrade," I said, "I'm going to send these home to the Peoria *Transcript*. I want them to tell the editor this war won't be ended until there is a total failure of the acorn crop. I want the folks at home to know what manner of men we are fighting."

That was early in my experience as a soldier. I never changed my opinion of the cause of the Confederacy. I was more and more devoted to the Union as the war went on. But I never questioned the sincerity of the men in the Confederate ranks. I realized how dearly a man must love his own section who would fight for it on parched acorns. I wished that his love and patriotism had been broader, reaching from the Gulf to the Lakes—a love for the Union rather than for a state. But I understood him. I hated his attitude toward the Union as much as ever, but I admired the man. And after Corinth I never could get a prisoner half-way to the rear and have anything left in my haversack.

Oh, I too have suffered the pangs of hunger for my dear country, as all soldiers have done now and then. But not as that Confederate soldier did. We went hungry at times, when rain and mud or the interference of the enemy detained the supply trains. But that man half-starved. That's different. After the battle of Nashville, December, 1864, we marched in pursuit of Hood as far as the Tennessee River. There, for more than a week, we subsisted on corn—not canned corn and not even popcorn, but common, yellow, field corn on the cob. And the row we suffering hero-martyrs made about it! A soldier was carrying a couple of ears of corn to a camp-fire to parch for his supper. A mule tethered near by saw him and lifted up its dreadful voice in piteous braying. The indignant warrior smote him in the jaw, crying, "You get nine pounds a day and I get only five, you long-eared glutton, and now you want half of mine!"

Bare Feet and Empty Stomachs

In those days of sore distress we learned various ways of preparing field corn to make it edible. We parched it and carried it around in our pockets, munching it at all hours and coughing the hulls out of our esophagi with raucous hacks. We made hominy of it, as the negroes taught us, boiling it in lye made from our abundance of wood ashes, and hulling it in mortars hollowed in the oak stumps. Then we learned to make corn pudding. This was hominy served on another plate for dessert. The "other plate" we obtained by scouring the same one with ashes and a corn-cob. Also we made corn pie, molding cold hominy into pie-like shape, very like the sauce of John Baptist Cavaletto in Marseilles prison who made what he would of his three hunks of dry black bread by cutting them into the desired forms of melon, omelet and Lyons sausage. So we made the hominy into the likeness of the dishes we "honed for." We used to say that we dined with the mules because the cook was on holiday.

Other haversacks we found that night on Corinth field with scant rations in them. Sometimes it was a chunk of corn pone. I used to think hardtack filled the order for concrete breakfast slab, but corn pone a week old reconciled me to soft food. Hardtack for mine.

So the southern people loved the states for which they suffered. As Professor Sloan writes of the French nation: "No people ever made such sacrifices for liberty as the French had made. Through years of famine they had starved with a grim determination, and the leanness of their race was a byword for more than a generation." That was why they held Europe at bay in their bare feet and with empty stomachs. Any cause for which we suffer deeply grows dearer to us with the suffering. We love it highly and holily. And when I listen to this beloved country of ours talking morning, noon and night about money, and money, and more money, I think of the parched acorns I found in the haversack of that brave Confederate soldier lying on Corinth field with his face turned toward the stars.

XII

GETTING ACQUAINTED WITH GRANT

SPEAKING of fighting and starving and eating,—you know you can't talk about war without discussing food, for soldiers eat a great deal more than they fight,—at a reunion banquet once upon a time a private soldier was called on for a speech. He rose to his feet a little nervously, looked up and down the crowded banquet board, and said:

"Boys, I am not much of an orator, but I will say this: There's a heap sight more of you here to-night than I ever saw in a fight."

And he sank into his chair overwhelmed by an avalanche of appreciative and enthusiastic applause.

I think one of the most impressive services held in the Federal armies during the war must have been a certain Thanksgiving sermon preached by Chaplain H. Clay Trumbull. He told me about it in one of the little talks I had with him which are fragrant memories in afternoon days. The lines in Virginia were drawn very close together, as so much of the time they were. It was stormy equinoctial weather, impassable Virginia roads had for days delayed the supply train, and officers and men alike were living on very short rations. But Thanksgiving morning a commissary train arrived, bringing to the Union soldiers plenty of government rations, and to one particular regiment came with providential timeliness good things from home —a veritable Thanksgiving feast. The pickets were firing spiteful little shots at each other as occasion served; occasionally a little skirmish marred the pleasure of the occasion; at intervals a field-gun boomed its defiance through an embrasured bastion. The blue smoke hung sullenly over the lines, and no soldier could know at what moment the quarrelsome skirmishing or a general's opportunity might bring on a battle. But the happy regiment, grateful for its anticipated Thanksgiving dinner of "mother's good things," mustered for service, and Chaplain Trumbull, inspired by the occasion and his always happy recognition of the only text for the special hour, preached from the passage in the Shepherd Psalm:

"Thou preparest a table before me in the presence of mine enemies."

A Homesick Warrior with No Place to Weep

What a meeting-house! What a congregation! What a text! And what a preacher! There was one occasion certainly in which the entire sermon was in the text. The soldiers were electrified with the wondrous harmony and appropriateness of the service and its environment. No man who heard it ever forgot the exposition of that passage. That was one of Chaplain Trumbull's rare gifts—recognizing a text that would preach itself. A good hint for a Sunday-school teacher. A good topic will illuminate a lesson as a headlight displays the track. One of the most delightful of American humorists—Charles Heber Clarke, of Philadelphia,—once said to me, "I wrote the book in about six weeks. Then I spent three months thinking of a title. Then the title sold the book."

This, however, is somewhat irrelevant. But not altogether so. Good things from home were more welcome to the soldier than December sunshine. I think that was especially so of the western troops. We wandered so far from home in search of our enemies sometimes. Texas was a long way from Wisconsin. It was a far cry from Michigan to Alabama. There was neither daily nor weekly mail. Letters reached us when and where they could catch up with us. Once upon a time, when I had been away from home nearly two years, our division was taken north by steamboats from Memphis to escort "Pap Price" out of Missouri, where he was having too much his own way with the state militia. And one day from Cairo to St. Louis we steamed up along the pleasant panorama of the Illinois shore of the Mississippi. I think there was also a shore on the Missouri side—there is now, I know, and it is quite probable there may have been a bank in that direction in 1864. I never saw it. I sat on the starboard wheel-house and saw every mile of that blessed Prairie State from Cairo to East St. Louis. That was Grand Tower, and that was Chester. There was the mouth of the Kaskaskia River, and there—oh, there was the "old Sam Gaty," headed for Peoria and La-Salle, and my mother wasn't a day away from me!

And I couldn't even go ashore—we never once touched at an Illinois port. Not once. And that old steamboat we were on—the *Des Moines*—it was the division "flag-ship" and was crowded from pike-staff to rudder with infantry and artillerymen. I prowled all over it, from pilot-house to hold, and there wasn't one secluded spot on that illy-contrived craft wherein a roystering warrior, who was at that time in the mounted service and an orderly at division headquarters—a swaggering trooper who wore clanking spurs and a jangling saber, tilted his hat far to starboard and made as much noise when he crossed the deck as a load of scrap-iron on a Philadelphia cobble street—there wasn't one place, I say, where that sort of a "mighty man" could go and have a good cry in any kind of comfortable privacy. So I saved my weeping until we caught up with old "Pap Price," and he gave me something that occupied my thoughts to the exclusion of tears. I have seen homesick people three thousand miles away from home. But oh, that trooper whose hat was pulled down over his eyes and who held his under lip with his hand to keep it from flopping against his teeth, was the homesickest thing that ever sopped his yellow-braided cuffs with tears because he was afraid somebody might see him if he used his handkerchief. But there is no comfort in that sort of a cry, when you are continually looking around to see if anybody is laughing. There is no pleasure in that kind of weeping. Every army transport that is built to carry young soldiers ought to be constructed with crying places.

Only a little way above Ste. Genevieve another boat in the fleet ran into us, smashed our wheelhouse and swept half a dozen mules overboard. All but one of them drowned. That happy hybrid swam ashore and landed in Illinois. He waved his paint-brush tail triumphantly and disappeared in the willows headed for Evansville on the Kaskaskia, right in the heart of Randolph County. The only mule I ever envied.

A box of good things from home—the only one that ever reached me during the war—was the simple means of introducing me to my commanding general. We were camped at Young's Point, Louisiana, where we were employed in digging that famous canal that was designed to carry the fleet around Vicksburg in that great campaign, when a man of my company came up from the river one day and said, "There's a box addressed to you down on one of the steamboats."

The Private Meets the General

While he was yet speaking, it seems to me, I had ascertained the name of the boat, got a pass to the river and an order for my box and was on my way. I presented my order to a civilian commissioner on the boat and was informed that all the stores on the transport, private and public, were the property of the Sanitary Commission, having been seized for use in the hospitals. Get into the hospital, he said kindly, and I could have some of the contents of my box. I said there was a smallpox hospital a few miles up the river that I could get into, but I didn't want the box so badly as that, although I did want it at almost any price short of the pest-house. I prowled around until I found my precious box. I showed it to the commissioner, feeling pretty certain that if it looked as good to him as it did to me he would relent and let me have it. The very sight of it produced in me a spasm of the same kind of homesickness I afterward contracted going up the Mississippi to St. Louis. He said he knew how tempting it looked, but duty—he paused impressively on the word and bade me remember what duty meant to a soldier. He pronounced it "d-double o." I lost my temper and told him I had heard the colonel say it much better and far more emphatically.

I tried one more appeal. I knew there would be letters in the box. Might I open it and get my letters? I had made a mistake in being sarcastic, and he wouldn't even let me do that, and finally ordered me off the boat. I think my lip must have hung down very pathetically, for the big Irish mate followed me to the gangplank.

"Ye'll get yer box, me lad," he said, "if ye do as I tell ye. Go up on the cabin deck an' ask the Ould Man."

Who was the Old Man?

"Ould Grant, no less. He kem aboard about an hour ago, an' he's up there smokin' this minute whin I kem down. I'll pass ye the gyard and ye'll go on up. Come an wid ye." He led me up to the cabin deck. There sat the silent brown-bearded man whose features every soldier knew and whose greatness every western soldier held in unquestioning reverence. I saluted, the mate explained my errand, and the general looked out over the turbid Mississippi and smoked silently while I pleaded my little case. Then he asked for my order. My heart beat high with the hope that he would write a military O. K. across it with magic initials. To my amazement, he read it and rose to his feet. "Come with me," he said. And a bewildered private soldier, escorted by the General Commanding the Military Division of the Mississippi, followed him to the civilian commissioner. I pointed out my property, and General Grant handed the order to the civilian.

The Inside of a Box from Mother

"Give the boy his box," he said simply. The commissioner bowed and I saluted. I wish I could imitate that salute now. It was a combination of reverence, admiration, kotow and renewed assurance of a distinguished consideration. Except possibly in China, the general never again received such an all-comprehensive obeisance. The cigar between the fingers swept a half-circle of smoke as the Commander, with military punctiliousness, returned the private's salute, and with a half-smile playing under the brown mustache, created, I fear, by that all-comprehensive, unprecedented salute of mine, he returned to his chair on the cabin deck, while the big mate patted my back all the way to the gangplank. That's why I love an Irishman.

And I? I simply unfolded the hidden wings which we wear on our feet for such occasions, and with a box as big as a field-desk on my shoulders flew airily and swiftly to the bower—that's what it was—tents of brushwood—of C Company, where I held high wassail with my comrades while we scraped that box to the bones. First thing, cake; then canned things and more cake; desiccated things; other kind of cake; condensed milk; layer-cake; can of preserves; sponge cake; jar of spiced things; fruit cake; socks and handkerchiefs; card of gingerbread; assorted things; perfumed soap; jelly; cookies; pocket-knife and marble-cake. The rest of the box was filled with cake. Mother knew what was good for her boy, all right. Didn't she raise him? She couldn't tell what physical or psychical changes being a warrior might have made in him, but she knew that the stomach he took away from home with him would last during the war. Only the remarkable fact that other mothers' boys have about the same general kind of stomachs saved me from the hospital that day. And just as we scraped the last cake crumbs together, all at once, with a unanimous community of sentiment, we remembered the captain.

Often as I journey to New York I have time to go out to the stately mausoleum on Riverside Drive, bearing over its portals the message of the great captain to the warring world—"Let us have peace." I stand uncovered as I look at the sarcophagus that holds his dust. I think of his greatness and of his simplicity. The courage of the soldier, the rare abilities of the general, and the gentleness of the man. I see him going with a private soldier, and hear him, in the voice that could have moved armies of half-a-million men, issuing the quiet command that gave to a boy a little box of things from mother. And that picture harmonizes perfectly with all the others.

XIII

WAR THE DESTROYER

"HELL's only two miles ahead of you," shouted the cavalryman with the voice of a prophet, mounted on a foam-flecked horse, black as midnight. He thundered down the column in a whirlwind of yellow dust, stormed with our cheers, for like an echo to his words we heard the dull "boom-boom" of a distant battery, and we caught the battle madness with the dust cast up like the smoke of an incantation by those flying hoofs.

Colonel McClure flung his arms apart in a gesture of command, and with cheers yet more deafening and hearts beating high with anticipation, the column broke with orderly disorder as we sprang to the preliminary work of destruction. For a battle always begins with destruction, before ever a shot is fired.

The colonel's gesture, clearly understood when

his voice could not be heard, sent us like human cyclones leaping at the fences that hemmed the road. Such a beautiful country we were marching through, that summer day. A park for loveliness; a granary for fertility. Low hills whose wooded crests smiled on the corn-fields that ran down to the emerald meadows. A creek meandering across the plantations, loitering in its broad and shallow bends to photograph the white clouds posing against the soft turquoise skies; stately old plantation homes with their colonial architecture; the little villages of negro quarters in the rear; pleasant orchards and fragrant gardens.

How beautiful they were, those sweet old southern homes! And dear and fair some of them still stand, here and there in the new South, amid the rush and clatter of modernity and progress, of steam and electricity, gasoline, automobiles and airships, tourists and promoters and prospectors, iron furnaces and coal-mines. Not as scolding protests against progress, development and prosperity—they are too gentle for that. They stand rather as beautiful memories of all that was sweetest and fairest and best in the Old South. What colonial grace in their white-columned verandas. What stateliness in the heavy cornice; what welcome of hospitality in the spacious doors with their old-time "side-lights," and in the sunny smiles of the many-windowed front. The shadow of pathos rests upon them now, tenderly as the sun-kissed haze of Indian-summer days. They temper our nervous desire for "newness"; they correct our taste for architectural frenzies of many-gabled deformities and varicolored creosote "complexions." They are of the old order, which, like the Old Guard, dies, but never surrenders to modern changes. They stood here before the war. They have been deluged with woe. They have been baptized in sorrows, the bitterness and depth of which our northern homes never knew—can not know—please God, never will know. And some of their anguish have been the common sorrows of all homes in war-times—the heartache of bereaved motherhood; the agony of widowhood; the loneliness of the orphaned. The loving Father of us all has made the sorrow that is common a healing balm that makes holy and tender the bitterness of the cruel past. The kisses that rained on the faces of the dead have blossomed into the perfumed lilies of consolation for the living.

A June Day Cyclone

And framing all that picture that lay along the line of march that June day, joining and separating all the fields with their zigzag embroidery, picked out here and there with the greenery of wild vines, and stitching in the winding yellow roadway as though it were a dusty river, were the old rail fences, picturesque in weather-beaten grays with the artistic trimmings of clambering festoons of leaf and blossom. A moment before our souls were drinking in this beauty until a little ache of homesickness added the bitter-sweet to the esthetic draught. Then, as the wild shouting ended, far as the length of the column wound along that road, there wasn't a panel of fence to be seen. Not one. Months of cheery toil it had taken to fence that highway out and shut the green fields in with a legal fence "horse high, pig tight and bull strong." Now as we picked up our grounded muskets or took them from the "stack," we looked upon an open country. A cyclone could not have accomplished the destruction more completely.

The fences had been a protection to the young wheat and the growing corn. They were the defenders of hungry men and women, of little children, white and black, who would cry for bread but for these barriers against marauding foes. The crooked lines of the old rail fence wore the dignity of high office. But now they were in the way. When there is going to be a fight the first thing is to prepare the ring. And war demands not a pent-up little twenty-four foot, rope-enclosed space, but many square miles in which its champions may maneuver. Its mighty wrestlers—Life and Death—must have abundant room. You build a platform and you construct a ring for your ordinary prize-fighters and wrestlers. But when real soldiers are going to give an exhibition of real fighting with the bare hand, the cold steel and the hot shell, you first destroy the country over which they are to fight. You set fire to that dear old mansion—it would shelter sharpshooters. You brush away these protecting fences. They would impede the swift sweep of cavalry; they would detain a battery ten minutes, and lose a battle; they would throw a line of charging infantry into disorder.

Scientific Destruction—Even for the Crows!

When we saw the colonel's gesture, tired we were, thirsty we were, hungry, faint and breathing dust. But with the light-hearted glee of schoolboys we sprang at those fences—a man to a rail—and they were gone. Sometimes we merely opened the panels like gates, leaving the alternate corners standing in the re-entrant angles. And the next squirrel that came running along his accustomed highway would pause bewildered in his up and down career along a fence builded entirely of gaps. But if there was plenty of time—say ten, instead of five, minutes—down to the level came all the fence.

That's war. Destruction of innocent and useful things. Destruction of everything. When we tore up a railway, it wasn't enough to demolish it so that trains could not go over it. We burned the ties. But we made them destroyers of other things in their own fiery death. We builded orderly heaps of them—because war does not destroy like a blind storm that does not know how to destroy properly—war destroys scientifically. On top of the ties we laid the iron rails. The heat of the fire furnaced the rails to red-whiteness, and their own weight compelled them to suicide. They bent down in strangling humiliation. Or, if there was time, fifteen or twenty minutes longer, men seized the ends of the rails with improvised tongues of twisted saplings, ran the red center of the rail against a tree, and bent it around the oak in a glowing knot. The enemy could make a new rail in less time than he could straighten out that entanglement. That's the way war destroys. An axiom of war is to leave nothing behind which the enemy can possibly use. "The next crow that flies across Shenandoah valley," said Phil Sheridan, "will have to carry his rations with him." That valley was unsurpassed in all the world for beauty and fertility. But it was also a granary and depot of supplies for the Confederate armies in Virginia. And when Sheridan rode down from Winchester town he was going to war. And war is destruction.

Don't censure Sheridan. That was civilized war. It is easy enough to say "barbarous," "brutal," "savage." For with all its ameliorations it remains war. As long as Christian nations justify war, they justify everything that it is and everything that it does. There is no such thing as a Christian war. Genghis Khan waged war about as Richard Cœur de Leon did. The Crusades were nearly as cruel as the marches of Attila. The invader is more destructive because of his greater opportunities.

The old German word for war meant "confusion." An old English word for it was "worse," as though it was worse than the worst thing you could name. It gives mourning for joy; ashes for beauty; the spirit of heaviness for the garment of praise. Law, a thing most sacred to our high civilization, is dethroned; the Sabbath is despised; Mercy is buffeted; Pity is struck in the pleading, tear-stained face of her. If another man doesn't dress as you do, he is worthy of death. If you say to him "Shibboleth," and he replies "Shibboleth," drive your bayonet through him. They did that at the fords of Jordan, three thousand years ago, and we haven't improved very much on the principle. That's war.

The Pitiless Wreck of Money and Men

War destroys everything. At one time it was costing the United States a million dollars a day to fight for its life. And what became of the million dollars? Destroyed. Burned up and broken to pieces. Gunpowder, wagons, cannon, tents, guns, drums, clothing. Burned to ashes, ground to dust; trampled in the mud; thrown into the river.

The broken musket is not mended; it is smashed against a tree to make the slight injury complete destruction. If the soldier's overcoat is a burden, he first tears it to pieces before he throws it away. The overturned cannon is abandoned; the broken-down wagon is burned; the lame mule is turned out to starve; the wounded horse is left to die in lingering agony—there isn't even time to shoot him. The injured arm or wounded leg that would be saved at home is amputated in rough haste. War can't even take care of its heroes properly. In the terror of defeat the wounded are left moaning on the field at the mercy of the night, the storm and the enemy. The hospital that tries to care for the sick and wounded feeds the grave much more than does the battle-field.

Even when it seems to spare, war destroys. A man's right arm is torn away at the elbow by a shattering fragment of shell. He is only twenty years old. And as they carry him back to the field hospital he thinks of the long years of life stretching out before him. Another young soldier lies on the operating table, and with set teeth and grim visage watches an attendant carry his amputated legs away to common burial with the ghastly débris of the hospital tent. A cripple for life—a helpless burden. And he is a farmer! A surgeon bends over another man to say cheerfully in cheery tones of encouragement: "You had the closest call a man could have and not answer it. But you're all right; you'll live!"

But the soldier knows that he will live in darkness, for the bullet that spared his life when it swept across his face put out its light forever. He'll never be the man he was before. War has destroyed him. Even the tender mercies of war are cruel.

Oh, I have seen war breaking men to pieces in this brutal fashion, as I have seen you with your switching cane behead the daisies laughing up into your face beside the meadow path. I have seen a soldier rise from a piano in a burning house, where he had been singing *Mother Kissed Me in My Dream* till our hearts were tender, and smash the ivory keys, blessed by the caressing touches of some woman's tender hands, with the butt of his musket. Why? Just to smash them. That's the way the war spirit transforms the hearts of men,—good, gentle-hearted men like your father, who was in my company; like David, who, in the sweet sunshine and shadows of the quiet sheep pastures sang, "Surely goodness and mercy shall follow me all the days of my life," and then in war time massacred the people of Rabbah, torturing "them under saws, and under harrows of iron, and axes of iron." That's war.

XIV

THE COLONEL

WHENEVER I think of him, there comes into my memory the lines of Guy McMaster in *The Old Continentals:*

> "Then the old-fashioned colonel
> Galloped through the white infernal
> Powder cloud;
> And his broad sword was swinging
> And his brazen throat was ringing
> Trumpet loud!
> Then the blue
> Bullets flew,
> And the trooper-jackets redden
> At the touch of the leaden
> Rifle breath;
> And rounder, rounder, rounder,
> Roared the iron six-pounder
> Breathing death!"

That was "the old-fashioned colonel." He may not have been especially scientific, but he was a terrific fighter, and after all, if fighting isn't the science, it is the business of war.

Because the Forty-seventh was a fighting regiment, it marched and fought, first and last, under five colonels,—all of them "old-fashioned." John Bryner, our first colonel, who marched away from Peoria with us in 1861; he died in the service, being reappointed colonel of the reorganized regiment in 1865; William A. Thrush, killed at the head of his regiment at the battle of Corinth, October 3, 1862; John N. Cromwell, killed at Jackson, Mississippi, May 16, 1863, our boy colonel; John Dickson McClure, wounded nigh to death in the siege of Vicksburg, June 20, 1863; Daniel L. Miles, lieutenant-colonel, killed in the battle of Farmington, Mississippi; David W. Magee, colonel in 1865. My colonels rode close up to the firing-line.

The relation of the colonel to his regiment was not merely that of a military commander. In the days of which I write, at least, it was paternal. He was the father of the regiment. Our most affectionate title for him was the "Old Man." Youth could not save him from this if we loved him. He did not receive this mark of honor and affection until we had tried him out for a few weeks, and had been at least once under fire with him. Then, if we decided that he would do, we began calling him "the old man," in much the same intonation of affectionate confidence with which a boy calls his father "daddy."

When We Had a "Regular"

I suppose there are boys who never call their paternal parent anything but father,—boys who would be whipped for calling him daddy. I always feel sorry for the father. And there have been colonels who would not tolerate the familiarity of "the old man." But I think they were colonels of militia. I never knew a fighting colonel who didn't like it.

We once had the honor of being commanded by a "regular." And it was an honor. Our regular colonel was Captain George A. Williams, whose battery of big guns the regiment supported in the battle of Corinth, where the captain won his majority for "gallant and meritorious service." He was appointed to command of the Forty-seventh in November, 1862, because our old colonel, Miles Thrush, was dead, shot through the heart at Corinth, in front of Williams' guns, and our new colonel, Cromwell, was a prisoner of war, captured at Iuka. We were "good boys," very fond of having our own way, and for some reason General Grant seemed to think that the fatherly discipline of a West Pointer would be good for our morale, and therefore appointed Major Williams.

We liked the "regular," who had been appointed to the Academy from New York in 1852, and was retired a colonel, I think, in 1870. We called him "the old man" after three days' service under him. The way of it was this. It was always desirable to "try out" a new colonel before he got too firmly seated in the saddle. One of the men of my own company detailed himself to trot a trial heat with the West Pointer just to find out what there was in the colt. He refused to go on a certain detail ordered by the sergeant, adding to his curt refusal that it took a bigger man than himself to make him do what he didn't want to do. As "Jacky" was the kind of man whose fists were in active accord with his word, the rather prudent sergeant who happened to be on duty that morning referred the soldier's insubordination to the company commander. This officer, who knew how Jacky had scandalized the company on one or two similar occasions by surrounding the entire non-commissioned force, ordered the sergeant to convey his prisoner to the new colonel.

Colonel Williams was a handsome, soldierly appearing man, with a smile that was as alluring as it was deceiving. He looked pleasant when the sergeant preferred his charge, and the prisoner promptly confirmed it, saying that the detail assigned him looked too much like work, and he didn't enlist to work.

Jacky Allowed to Play

The smile on the face of the colonel brightened.

"No," he said, that was true. No soldier liked to work. He was not overly fond of work himself, although he was sometimes compelled to do very hard, disagreeable things. "So," he concluded, "since you dislike work, you shall play all day."

Under instructions, the sergeant marked out a circle on the parade ground about twenty feet in diameter. The soldier who didn't like to work was given a log of fire-wood about six feet long and heavier that a knapsack full of stones. Guards were set and regularly relieved, and the prisoner began his play-day by walking around that circle with his burden. All day long, from guard mount in the morning to dress parade in the evening, he lugged that load of fuel. The guards, who now stood in terror of this new sort of good-natured colonel who wouldn't stand the least bit of any sort of foolishness, and who smiled like a seraph when he put a man on the treadmill, were afraid to permit the prisoner to halt when he ate his dinner, which the cook brought to him on a tin plate. The sentry allowed him to lay down his burden while he ate, "But," he said, "you've got to keep walkin' !" And walk he did, wearily shifting the log from aching shoulder to aching shoulder. He was released after dress parade.

Instead of throwing the heavy log down gladly and indignantly, he stooped and laid it on the ground as though it was a sleeping baby. "If I had thrown it down as hard as I wanted," he afterward explained, "I would have broken it into half a dozen pieces, and there's no telling what the old man"—he had learned that during his march—"would do with me then. . . . I reckon," he said as he rolled into his blanket that night, "that I toted that fuel train twenty miles to-day, and never got half a mile from where I started."

Colonel Williams was christened "the old man" that day. We liked him immensely. He knew how to get things for the regiment that volunteer officers didn't know how to ask for. "The old man," said Corporal Lapham, "knows how to get things the other colonels don't know the Government's got." He made us dress better, stand better, keep neater, behave more soldierly and jump more promptly at an order. He fed us better, got more new uniforms and blankets for us, stocked the hospital with more and better supplies. We liked him, we obeyed him, we were just a little bit afraid of him, and we were genuinely sorry when he went back to his own command. Soldiers do love a colonel with a bite right close behind his bark. Why else should he be a colonel? "I like the rooster," said that quaint old philosopher, Josh Billings, "for two things: for the crow that is in him, and for the spurs he wears to back up the crow with." A crow without spurs is a blank cartridge.

Because of this paternal responsibility with which the men invested him, the colonel of a volunteer regiment was burdened with a hundred and one things that should never have reached him. We went to him with complaints that should have stopped short at the sergeants, or at the furthest never passed beyond the captain. But we felt that we had a right to see "the old man" about everything. As a rule he listened to us, although more than half our wrongs were imaginary, and the other half of our hardships were either richly deserved or inseparable from the soldier's life. Happy the colonel with a sense of humor to sit on the judgment seat beside him. Once upon a time we sent a delegation of three men to complain of the fearful quality of the company cooking.

The Committee's Unexpected Success

The colonel agreed with us without tasting the samples of food we brought along. "Yes," he said, "you have a wretched cook. I am going to detail him to cook at regimental headquarters, where I can watch him. I'll transfer him immediately."

The committee came back with faces of consternation and reported. A roar of indignant remonstrance went up from the assembled rank and file. "What! Take away Billy Wanser! Take away the only man in the regiment who knew how to cook? The only man who never had a meal late? The man who caught up with the company in the rain and mud ahead of the supply trains? The only cook in the regiment who came out on the battle-field with hot coffee? Take away Billy Wanser? Not over our dead bodies!" And we hastened to Colonel McClure's headquarters in a uniformed mob to denounce the unfortunate delegation whom we had sent there half an hour before, as a self-appointed squad of malcontents who deserved to be starved. And the colonel agreed with us and said he would put in irons the next man who dared slander our matchless cook.

Of all the colonels under whom I served, Colonel John D. McClure was my ideal. A man with a strong figure and a strong face, a man's voice, deep and commanding; clear steady eyes, that shone with the kindliest glow that ever turned into a steely gleam when they looked through the shuffling excuses of a skulker. He was captain of Company C when I enlisted. When he reached the colonelcy by successive merited promotions, the men of C Company called him "the old man" before he put on his new uniform. He was as kind-hearted with his men as a good teacher is with children. If a question of discipline trembled uncertainly in the balances, mercy always tipped the scale with a gentle touch of her lightest finger —but it was enough. He was at the side of a sick or wounded soldier as quickly as surgeon or chaplain could reach the sufferer, and there was encouragement and consolation in the deep voice of the colonel. Under fire, his calmness was contagious. His courage rose above excitement. There was none of the hysteria of battle about him. He was never a "noisy" colonel, though his shouted orders reached every man in the regiment, and "his brazen throat rang trumpet loud" in leading line or column. I never heard him use a profane expression. He was gentle as he was brave; quiet as he was manly. The regiment loved him because he was of lovable quality. Once, while we were in quarters at La Grange, Tennessee, his young wife came down to the front to see her husband. Virginia Cunningham—as sweet and beautiful as her husband was noble. She and I had been schoolmates in the Peoria High School. And if in those days of childhood dreamings I had ever prophesied that one day she would marry my colonel and thereby share his authority to say to me "Come" and "Go," we would have laughed over it as the merriest bit of fiction an unbridled imagination could devise for a summer day's fooling. But that was just what happened.

Gentle he was, and kind-hearted. But we all knew there was but one law in the regiment. That was the colonel's word. It was quietly spoken, as was his way, in counsel or on the field. There was no fulmineous profanity to emphasize it and no Jacksonian appeals to heaven to confirm it. But it was respected. His quietness magnified his firmness and courage. His "gentleness made him obeyed."

XV

A TRIPTYCH OF THE SIXTIES

"Ready!"

ONE June morning in 1863 we were ordered to report at Fort Pillow in parade accouterments to see three men shot.

Here was a novelty in our military experience. We expected to see men shot every time we went out. We had seen them shot by scores and hundreds in the sharpshooting of the skirmish line and in the fearful volleying of the line of battle, by musket and by artillery, but never before had we received instructions to march out and, with empty muskets, to form in the square of parade and witness the official shooting of three of our comrades.

This was a new kind of killing. I know that not a man in my regiment had ever witnessed a military execution. I doubt very much if a soldier in the entire division had ever seen such a thing. There was a chill in it that doesn't come with the ordinary death of a soldier. We knew there were crimes against military discipline and soldierly righteousness that were punishable by death, but it had never come near to us—never so near as this. I had seen men punished before the regiment for the crime of desertion. I had watched them while the corporal cut the brass buttons from their uniforms to destroy, as far as possible, every vestige of the soldierly uniform the culprit had disgraced. Then I watched them shave one-half his head down to the white and glistening scalp, and so tragic was the picture that it did not look grotesque; though ordinarily one would laugh at such a thing. I had heard his sentence read; then, as the fife shrilled and the drums played the lively measures of the *Rogue's March,* the culprit followed the corporal's guard down the front of the line. Here, at the right of it, he halted, saluted his colonel, and it was his duty to thank him for the just punishment that had been meted out to him.

In the only case in which I ever was a witness, however, the dishonorably discharged soldier yelled at the top of his lungs: "I am a civilian now! To hell with you and your shoulder-straps! I am as good a man as you are!" And the colonel had the grace not to add punishment to the punishment already inflicted on the man maddened by the sting of humiliation. The man disappeared from the camp that night. What became of him afterward I never knew.

But here was a case of capital punishment in that great organization of the army that was formed and drilled and trained to administer capital punishment to the enemies of the republic. Every man in a hostile uniform who leveled his musket at the Stars and Stripes adjudged himself guilty of treason and sentenced himself to death.

But all this in the heat of battle. This with the blood hot and the pulses throbbing and the stress of conflict knotting every muscle and stretching every nerve to a tension like a harp-string. What we were to witness this beautiful June morning in the suburbs of the busy city of Memphis was something entirely different.

Three men were to be punished to death for an offense not only against military discipline and soldierly good conduct, but for an offense recognized as gross under the civic law, an offense against civic righteousness and morality; an offense aggravated by the circumstances surrounding its commission. It seemed strange to me, as I put on my accouterments, that we were to shoot these men for an offense against the civil law for which the civil law provided no capital punishment.

The three condemned men had occupied one of the most responsible positions in which a soldier is ever placed. Just outside the city a few miles they were on picket duty. They had the keeping of the city's garrison and the surrounding camp in their care. They possessed authority equal to that of an officer. They were to scrutinize every person who came and went.

About mid-afternoon there came from the city

to the post of these three pickets a southern farmer, his wife and their grown daughter — a young woman. Their pass, signed by General Hurlbut, was correct, and they were told to pass through by the soldier who read it and the corporal who looked over his shoulder. It was simply an incident of the day. A few moments later, when these citizens had passed but a few miles farther on, the old man returned, having met with an accident. He had broken a wheel so badly that he could not repair it. He must go back into the city and secure the help of a wheelwright. When he had explained the conditions to the soldiers they gave their permission and he went back into the city, leaving his wife and daughter a few miles beyond the lines, but under the protection of this picket guard, than which they should have had no stronger protection. Then the devil got into the hearts of these men. Bad men they may have been—wicked men, base men, but they had been good soldiers else they had not been placed in the responsible position of picket guard on lines so close to the city. Through what lines of discussion they came to their cruel and foul decision I do not know. When a man palters with his duty he is always on the way to betray it. They took the first false step. They abandoned their duty and hurried out along the road where the disabled wagon was waiting the return of its owner. There upon the persons of these helpless women, confided absolutely to the protecting care and honor of these soldiers, whose wards they temporarily were, they committed a crime, to a woman worse than death, doubly horrible from the fact that the victims of the lust of these soldiers were mother and daughter.

In the course of an hour the man returned. The wife and daughter told him of the horror that had befallen them. Straightaway he walked back into the city, not asking permission this time, as he knew the soldiers would not dare to refuse it, and reported to the proper officers what had transpired on this picket post. An arresting party marched out, the three soldiers were taken off duty, disarmed, taken back to the guard-house and placed in irons to await the promptness of a military trial, a drumhead court martial. Their foul offense was set forth quickly and their guilt was proven. The unanimous decision of the court martial was that they were guilty of a crime punishable by death and they were sentenced with little delay.

Punctuality is a military virtue. We did not wait long in parade formation for the fearsome event of the morning. All about us were gathered, in the rear of the uniformed ranks, the motley mob of the city, white and black. Busy hucksters were plying their trade, hawking their wares of sandwiches, little cakes and coffee, seeking to make their profits at the gates of sudden death.

"*Aim!*"

There was a good deal of talk in the ranks as we stood at rest. The seriousness of the affair did not seem to oppress the men. They were not depressed. There was a unanimity of approval of the justice of the sentence. There was wrath in the tones in which many of the men condemned the offense of the culprits. All true soldiers felt the shame and disgrace that had stained the United States uniform. We felt that in their crime somehow they had smirched the rest of us.

The three men were cavalrymen—members, if I remember correctly at this late date, of the Third New Jersey Cavalry—and I can remember so well how so many of the men congratulated one another and themselves that the offense had been committed by an eastern regiment; for we insisted that the native chivalry common to the western men would have held them back from the commission of such a crime. We surmised that these men were not true types of the eastern soldier. From the purlieus of Jersey City, from the slums of Hoboken, from the overcrowded districts of Paterson, we said they had come. It was rather a shock to our satisfied philosophizing to remember afterward that one of these men was a farmer's boy. Vice does not mark its boundary lines by the streets of city wards or the lanes of the country.

One man, I remember, excused the farmer's lad.

Some man from the mighty wheat country of Minnesota wanted to know defiantly what you could expect of a man brought up to call a ten-acre Jersey truck patch a farm. And this excuse being better than none, we all agreed with him that the man's environment and training were bad.

There was a burst of military music—not the wail of the fifes, intoned by the monotonous roll of the muffled drums, touching the heartstrings with the thrill and pathos of the old, old dead march, with its plaintiff measures that had endeared itself to thousands of hearts on either side of the sea when, with honor and sorrow, we buried our dead who died like men, who died on duty—brave soldiers with clean hands and loyal hearts. It could not be degraded to such a service as this. But the band played mournful strains of a march that seemed to emphasize not only sorrow, but shame.

Came the band into the square, wheeling sharply to the right, and marched down the three sides of the open square. Upon the orders gruffly shouted by the colonels, barked by their men down the lines, with the rattle of musketry and the jingle of accouterments, the troop came to attention. Behind the band marched the guard; and we knew even then that they marched, not as at the ordinary funeral of a soldier with arms reversed, but with fixed bayonets and pieces at the "carry"—a guard on duty; men with stern faces, officers with rigid lips. Behind the guard rode the colonel commanding the regiment to which these men had belonged. Then came two battalions of their own regiment—the only troops of the organization then in Memphis. For our cavalry regiments were often cut up into small detachments doing scout service for infantry garrisons and columns. The men carried their carbines in the slings. Their sabers were drawn, leaning with glistening brightness against the shoulders of the troopers. The hoof-beats of the horses fell with a muffled sound on the turf, and the jingle of bit and spur carried a military accompaniment to the cadences of the band.

Following the regiment of the disgraced men came an army ambulance with the cover removed.

Sitting on the coffin in the ambulance was one of the condemned men; kneeling beside him, a chaplain. A second ambulance brought its burden of condemnation; and a third. Two of the men were Roman Catholics, and chaplains of their own faith attended them—priests of the church that never lets go of a man once she has held him in the faith of her communion, but to the very gates of death carries her assurance of pardon on repentance and confession, and grants the indulgence of sins forgiven through the plenary power and authority of the priest who represents that great church. The third man was a Protestant, and a chaplain from one of the infantry regiments knelt beside him.

Marching slowly down the sides of the parade, all voices now stilled, not only by the command of military discipline, but by the awe of the occasion, the death march led the way of the procession. As the condemned men passed near to us we saw how white and set the faces were. On the battle-field they would have met death with flushed faces and kindling eyes—the bearing of brave men. They were going to a dishonorable death, and shame covered every face as with a garment.

Under the walls of the fort, on the open side of the parade, three graves were dug, like trenches, in a little line. The men dismounted from the wagons, details of soldiers carried the coffins; each was deposited on the side of the grave farthest from the troop, between the edge of that awful cavern of darkness and that little mound of earth. Details of soldiers bound the feet and the arms of the men fast—the arms behind them—and they sat down, each on his own coffin, facing his comrades.

A lieutenant read in brief summary the story of the crime as brought out by the court martial. He read the finding of the court and the sentence, and, finally, the military order of the colonel commanding the execution, indorsed by the signature of the commanding general.

From the left of the condemned regiment now marched a detachment of thirty dismounted troopers with their sabers sheathed, holding their carbines at a carry. They were formed in front of the condemned men, standing between the graves and the troops—a detachment of ten men facing each prisoner. There was a final word from the chaplains to the men. The chaplains, retiring, took their places far to the left of the coffins, out of range of the firing party. In a voice so low that it could not reach us (we only knew what it was because we knew what it must be) the order was given, "Load at will." The cartridges were placed in the carbines.

There stood now between the condemned men and the soldiers just the firing party. Off to the right, and in front of them, stood the officer in charge. The execution was conducted in deathly silence. The officer drew a white handkerchief from the breast of his uniform jacket and held it in the air. We heard, following the gesture, the clickety click of thirty carbine hammers. With a dull thud the butts of the carbines were lifted against the shoulders of the firing party, for that gesture said "Aim!" The officer unclasped his fingers. The handkerchief fluttered to the ground like a beautiful snow-white butterfly, the Japanese emblem of the soul. With the sound of one musket the carbines rattled their deadly volley.

"Fire!"

I expected to see the men on the coffins leap to their death. One of them swayed, indeed, drunkenly, first to the left and then to the right, and fell on his side on his coffin. The second one bowed slowly forward, falling with his face on the ground. The third one swayed backward, as gently as though some unseen hand had pressed him, and lay with his feet across the lid of his casket, his head and body hidden from our view. A sergeant stepped briskly forward and stood for a moment stooping over the face of each man. He turned to the officer commanding the firing party and made his report. Each man had been struck—and fatally. The wages of sin had been paid.

A shouted command from the colonel of this

regiment, and that, and that, and with the staccato repetition of the command from the line officers, the troops wheeled into column. The drum corps of the regiments, taking the place of the now silent band, struck up a lively marching air and, timing our steps to some well-known marching tune, we were hurried back to our respective quarters.

I remember well, with a certain reckless, soldierly sense of grotesque suggestiveness, our own fifes and drums led us, half marching, half dancing, back to our parade-ground to the merry steps of *A Rocky Road to Dublin.* It was the reaction. It was the setting of the lesson. It was the moral of the true fable we had just witnessed—the inevitable "*Haec fabula docet*"; "This fable teaches." The band was chanting, in staccato measures and rollicking time, the proper "recessional," *Lest We Forget.* That was the tune. *A Rocky Road to Dublin* was the hymn. "The Way of the Transgressor Is Hard" was the collect. The service was ended.

XVI

THE FAREWELL VOLLEYS

WHAT happened next was this:

The orderly sergeant, Dexter M. Camp, came to me with his little book in his hand and said:

"Burdette, make yourself look neat and smart. You are detailed for funeral service, and will be one of the escort. Report at once to Corporal Davidson."

A funeral? I had been in the army then more than a year. I had helped to bury the dead on more than one battle-field. But I had never attended a funeral. I knew that my comrade was dead. And I knew of course that he would be buried that day. But it had never occurred to me that he would have a funeral.

When we buried Private John Taylor, of C Company, the Forty-seventh Regiment of Illinois Infantry, War, who slew him, demanded that we, whom he might also slay as opportunity offered, pay due and formal reverence to one of his dead. We should observe the ritual to the letter. Himself, in glittering helmet shadowed with sable plumes, would review the funeral procession, and give to the occasion the environment of pomp and glory which the dead man could never have won had he passed away in his quiet home in La Salle County.

So the sergeant detailed six pall-bearers, of the dead soldier's own rank, and an escort of eight privates under command of Corporal Davidson.

When the commanding general is buried, the minute guns boom their salute from sunrise until the march to the grave begins. Officers of high rank are selected for pall-bearers and escort.

When the colonel dies, his entire regiment follows its dead leader to his grave, even as it followed him to his death.

For the dead captain, his company marches as his escort.

And when we buried Private John Taylor, we

followed the "Regulations" in the detail of pall-bearers and escort. All the non-commissioned officers of the company were required to follow the detail, and when the commissioned officers attended the funeral, they marched in the inverse order of their rank—the escort, the privates, corporals, sergeants, lieutenants, and in the rear of all, the captain.

When the platoon was formed, the pall-bearers carried the body down in front of it. The corporal gave the order—

"Present!—arms!"

"*The Land o' the Leal*"

An honor never accorded the living private. You see, Death is a king. And when he holds high court, he ennobles the soldier upon whom he has set his signet of distinction. When the body of the dead colonel is carried before the regiment, the lieutenant-colonel gives this same order. The regiment pays to the colonel the same honor—no higher

—which the escort of eight men paid to Private Taylor. Death levels to the rank of the soldier all titles and grades of authority or nobility. "Dust to dust."

The pall-bearers, having halted to receive this honor to their burden, carried the body to the right of the line. Again the corporal's voice—

"Carry—arms! Platoon, left wheel—march!

"Reverse—arms! Forward, guide right—march!"

The dull flam of the muffled drums draped in crape gave our steps the time. Then the wailing fifes lifted the plaintive notes of the dead march, which was oftener than any other *The Land o' the Leal*, and the drums beat mournfully in the long roll with the cadences that emphasized its measures and moved our marching feet in the slow rhythm of the dirge.

Somehow the sunshine seemed dim and misty as the muffled drums spoke mournfully. Our slow steps seemed to be timed not only by the throbbing drums but by the heart-breaking sobs in a far-away northern home. The fifes filled the air with tears. The sweet voices of women, tremulous with sorrow, blended with the music of the march—

> "Ye were aye leal an' true, Jean,
> Your task's ended noo, Jean,
> And I'll welcome you
> In the land o' the Leal."

Women? He had not kissed a woman since he left the little home in Illinois. That was one thing that made the old Scotch melody ache with its plaintive wailing. When he was so patiently wrestling with death in the loneliness of his tent, a woman's voice speaking his name had sounded to him like the blessing of God. So much of love, and tenderness, and longing prayer, and ministering touches of gentle hands centers about a deathbed at one's home.

When a Soldier Dies

But the soldier? His last looks upon his kind on earth were bent upon bronzed or bearded faces. Hands that would minister to him in his growing weakness were hard and calloused with the toil of war, and scented with the odor of the cartridges they had handled. Kindly faces, yes; but they shrank, man fashion, from trying to look too sympathetic; the voices were hearty and frank and jovial. Men are awkward in the sick-room, and a soldier resents being "coddled." A soldier's death is one of the saddest things on earth.

To die, and know that in his home voices were laughing and hearts were light. They were talking merrily over some jesting line in the last letter from him; they were counting the months against his return; they were planning such singing festivities when he came home—

And it would be days, days, days before they would know that he had gone Home, and was waiting for them. What would be the measure of their sorrow, bereaved of the mementos of death for which we long? His dying kiss; his last spoken word, with its message of infinite tenderness and love; the name that last lingered in a whisper on his lips; the look that lighted up his face just before he closed his eyes—the peace that God lovingly printed on his tired face—these things they would never know. For even we, his comrades, who would have died for him, were not with him when he passed away. "We could not be with him" when, in the loneliness of a soldier's death, he passed from that little shelter tent into the splendor of the building that hath foundations.

These were the things with which our thoughts were busied as, with "arms reversed," we followed the throbbing drums and the wailing fifes. Our hearts were heavier than the burden on the bier, and but for the shame of a noble thing our tears had dropped fast as the beats of the muffled drums. The drums; in their sad monotones they seemed like the pattering of a woman's tears upon the coffin lid. They modulated the shrill grief of the complaining fifes, as the heavy voice of a man, tremulous with a common affliction, soothes the pleading anguish of a heart-broken woman.

That was forty-nine years ago. But I had the heart of a boy, sensitive to all impressions. And to-day I can feel the ache coming into my eyes when I hear the crying of the fifes and the sobbing of the muffled drums.

> "There's nae sorrow there, Jean,
> There's neither cauld nor care, Jean,
> The day is aye fair,
> In the Land o' the Leal."

We reach the grave. Wailing fifes and sobbing drums are silent. The platoon is halted.

"Right wheel into line—march! Carry—arms!"

The bearers bring the coffin down the front of the line, halting in the center. Again the corporal—

"Present—arms! Carry—arms!"

The coffin is rested beside the grave.

"Rest on arms!"

The muskets are reversed, the muzzles resting on the left foot; the hands of the soldier are crossed on the butt; the head is bowed on the hands; the right knee slightly bent.

The chaplain steps to the front and center. Here he is greater than the colonel. He reads from a Book—the only Book men read from at such a time. It is a soldier's Book. The first words of the chaplain ring out over that open grave like the glorious triumph of victorious bugles—the trumpets of the Conqueror—

"I am the Resurrection and the Life; he that believeth on me, though he die, yet shall he live; and whosoever liveth and believeth on me shall never die."

The Cry of Victory at the Grave

Splendid! Magnificent! Right soldierly! "In the midst of death we are in life!" That is the way to read it. Read on, brave chaplain! Oh, I never stand beside an open grave that I do not see the Son of God standing on the other side of that narrow chasm of shadows, in the resplendent beauty and glory of the perfect Life. I hear him calling across to us: "I am the Resurrection, and the Life!" "I am he that liveth and was dead, and behold I am alive forever more!" I would not qualify by the slightest shading my absolute belief in that glorious teaching of the Living Christ, I would not exchange one positive word of it, for the most perfect comprehension of all the cleverest guesses and most brilliant doubts of all the wisest scholars that all the ages have brought forth in this world of human theories and mistakes and re-statements, conjecture and hypothesis. I did not doubt it that day, with the bearers holding the dead man on their shoulders before me; I have never for one moment doubted it since. When I get to Heaven I will be no more certain of it than I am now.

Over the body of the dead soldier the chaplain lifts our souls in prayer to the Living God. He steps to his place at the right of the platoon. The corporal commands—

"Attention! Carry—arms! Load at will—load!"

The rattle of rammers and the clicking of the musket-locks.

"Ready—Aim—Fire!"

Thrice the salute is fired over the soldier's grave. The clouds of blue smoke, the incense of war, drift slowly skyward above the open grave, as though they might carry with them the soul of the dead, obeying the call of the resurrection.

"Carry—arms! By platoon, right wheel—march!"

"Forward, guide left—march!"

The somber crape has been removed from the mourning drums. The rattling snares are tightly stretched. Clear and shrill as lark-songs the merry fifes trill out the joyous measures of *A Rocky Road to Dublin;* the stirring drums put the tingle into our half-dancing toes and the spring into our heels; "Right shoulder shift 'em!" jocularly calls the corporal, and with laughter and chatter we march back to camp, and life, and joy, and duty, and death—"all in the three years!"

Had we, then, forgotten him so quickly? Forget the comrade who had shared our duties, our privations, our hardships, our perils? It was nearly fifty years ago that we fired our "farewell shot" over that grave, and a little ache creeps into my heart with the thought of him to-day.

It isn't a good thing for a soldier, who every day

must face death in some measure, to be depressed in spirit. It unfits him for his duties. The trilling fifes and the merry drums are not to make us forget. They are to remind us that we must be ready for every duty, cheery and brave and faithful. The music of the camp never dims the memory of the comrade who had been called to higher duty. It's the way of the camp, and of the busy world, and it's a good way. I do not believe in wearing mourning for the dead, yet no man loves his friends more dearly than I. I would not say of my loved ones, when they pass on to the perfect life, "They make me gloomy every time I think of them. As a token of my feelings toward them, I darken my sunshine with these sable garments of the night."

The origin of wearing mourning garments was not to express sorrow, or reverence. The peculiar garb was assumed to warn all persons that the wearer was "unclean" from contact with the dead, and was therefore to be avoided, as a leper is shunned.

"Fall in for roll-call! Attention!"

The living respond to their names. The sergeant calls the details. This man for camp-guard; this one for picket; this one to hew wood; that one to draw water; you to go forth to battle; you to minister to the hospital; you to abide in your tent, waiting to be called.

And the dead, having achieved their full duty, sleep sweetly and quietly, waiting to hear Him say, "I am the Resurrection, and the Life."

XVII

THE LOST FORT

Morning

SILENCE, and the darkness before the dawn. Across the meadows, through fields of trampled grain, and far down the aisles of the forest, the stacked muskets mark the multiplied lines of the bivouac, broken here and there by the dark squares where the batteries are parked. Along all the lines the camp-fires smolder in their ashes. Across the velvet blackness of the sky the starry battalions march in the stately order of a million years—squadrons of the glory of God. Now and then, as a bearded veteran might lightly and smilingly touch the shoulder of a little child, playing at war, proud of his toy gun and paper epaulet, a great star that has flamed the splendor of the Almighty since time began, touches with a flash of golden light the bayonet of a sentinel, guarding the slumbers of his wearied comrades. Tired as the weariest of them, his own eyes burn and his body aches for sleep, but Honor on his right side and Fidelity on his left wind their mighty arms about him and keep pace with his steady step as he walks his beat. He is but a man and he may go mad from sleeplessness; but he is a soldier, and he will not sleep. The morning darkness deepens. It gathers the sleeping army into its silent shadows as though to smother it in gloom.

Into the silence and the night, as a star falling into an abyss, clear, shrill, cheery, insistent, a single bugle sings, like a glad prophecy of morning and light and life, the rippling notes of the reveille. Like an electric thrill the laughing ecstasy runs through all the sleeping, slumbering ranks. A score of regiments catch up the refrain, and all the bugles—infantry, battery and flanking troopers—carol the symphony to the morning. Shouting and crowing soldiers swell the chorus with polyphonic augmentation; the shrill tenors of neighing chargers answer the "sounding of the trumpets, the thunder of the captains and the shouting." And from all the corrals of the baggage and ammunition trains, the much-derided mule, equally important and essential in the success of the campaign as his aristocratic half-brother, raises his staccato baritone in antiphonal response. The camp, that a moment since, lay in such stillness as wrapped the ranks of Sennacherib when the Death Angel breathed on the face of the sleeper, is awake. And if one closed his eyes to shut out the gleaming bayonets and the stacked muskets, and the guns, silent and grim, muzzled by their black tampions, and only listened, he might think he was in the midst of a mob of joyous, care-free, happy schoolboys out on a vacation lark. For a soldier is a man with a boy's heart. The heart of the morning on the march sings in the notes of the reveille—joyous, free, exultant; it is the very ecstasy of life; the thrill of strength; the glad sense of fearlessness and confidence; a champion's desire to match his strength against the courage and prowess of a man worth while. On every camp of true-hearted soldiers rises "the sun of Austerlitz."

Noon

Straight over the earth hangs the great blazing sun, as though he poised in his onward flight for just a second, to say, "I want to see the very beginning of it." He flames down on the long trail of yellow dust that stifles the marching columns. The songs are hushed, for the feet are tired and the throats are parched. The fours are straggled across the roads, as the files find the easiest path for the route step. Conversation is monosyllabic. A soldier barks out a jest with a sting in it, and catches a snarl in response. A tired man, with a face growing white under the bronze, shakes his canteen at his ear, and decides that he isn't thirsty enough yet. A trooper comes galloping from the front with the official envelope sheathed underneath his belt, and is joyously sung and shouted on his way along the rough edges of the road by the sarcastic infantrymen, momentarily grateful for the diversion of his appearance—a human target against which all their shafts of wit and taunt can be launched, with the envy of the soldier with two legs in his hereditary jousting with the one who glories in six. The trooper is gone. "The tumult and the shouting dies." Again the long winding road; the yellow dust; the hills, the blazing sun; the cloudless sky; the tired men; the silent impatience over the step that has been quickened apparently without orders; the long stretch of marching since the last rest; an occasional order barked by a line officer, to correct the too disordered formation; over all, the hot stillness of noon. The morning breezes died long ago. The air is dead. The leaves on the forest trees that line the road swooned with the prayer for rain in their last faint whisper to the dying zephyr that kissed them in its passing. The dust of mortality covers their brave greenery —the same yellow dust that veils the phantom army marching past.

So far away—away in the advance, and far on another road—so faint and dull that it scarcely seems to be a sound, but rather a sensation that runs past the unguarded portal of the ear to touch the brain—the echo of a dream—Boom!

And yet it is deadly clear; fearfully near. Every listless head in the weary ranks is lifted. Questioning eyes answer one another. Every soldier has read the message, shouted so far away by a tongue of flame between black lips. Unconsciously the marching ranks are locked. Instinctively the step is quickened. The man with the whitening face drains his canteen to the last precious drop. He is going to have strength to get to the front with the regiment. Then, if he dies, he will die in the line. "Chuck-a-chuck!" the very battery wheels put a defiant tone in the old monotony of their rumbling. "Clippity-clippity!" another galloping trooper goes down the column in a cloud of dust, but this one is garlanded with cheers, and his face lights with a grim smile. "You'll find somebody that'll make you holler when you ketch up with the cavalry!" floats over his shoulder. "It's his deal," laughs a soldier, pulling his belt a buckle-hole tighter. Tramp, tramp, tramp.

A single rifle-shot. Sharp; penetrating; anger and surprise in its defiant intonation. A score of excited echoes clattering after it from hill and forest. A thrill of nervous tension runs through the column that closes the ranks in orderly formation. Quick terse orders. Absolute discipline in every movement. The crooked rail fences on either side the road are leveled as by magic as the hands of the men touch them. The column double-quicks out of the road to right and left. Curtaining woods swallow it. The men drop on their faces. They are lost from sight. The skirmishers, deploying as they run, swarm down the hill slope to the front like a nest of angry hornets. A handful of shots thrown into the air. They have found the pickets. A fitful rain of skirmish firing; a shot here; a half dozen; a score; silence; another half-dozen shots; a cheer and a volley; far away; ringing in clear and close; drifting away almost out of hearing; off to the right; swinging back to the left; coming in nearer; more of them, gathering in numbers and increasing in their intensity; batteries feeling the woods; a long roll of musketry; ringing cheers; thunders of awakening field-guns on right and left; the line leaps to its feet and rushes with fixed bayonets to meet the on-coming charge; the yellow clouds have changed to blue and gray; sheafs of fire gleaming through the trees; sickles of death gathering in the bloody harvest; yells of defiance and screams of agony; shouting of "the old-fashioned colonels" who ride with their men; bayonets gleaming about the smoke-grimed muzzles of the guns; fighting men swarming like locusts into the embrasures; saber and bayonet, sponge staff and rammer, lunge, thrust, cut and crashing blow; men driven out of the embrasures and over the parapet like dogs before lions; turning again with yelp and snarl, and slashing their way back again like fighting bulldogs, holding every inch they gain; hand to throat and knife to heart; hurrying reinforcements from all sides racing to the crater of smoke and flame; a long wild cheer, swelling in fierce exultant cadences, over and over and over the reversed guns, like the hounds of Acteon, baying at the heels and rending the bodies of the masters for whom but late they fought; a white flag fluttering like a frightened dove amidst smoke and flame, the fury and anguish, the hate and terror, the madness and death of the hell of passion raging over the sodden earth—the fort is ours. *Io Triumphe!*

Night

Count the dead. Number the hearthstones, whereon the flickering home-light, golden with children's fancies and women's dreams, have been quenched in agony, heartache and blood. Take census of the widows and orphans. Measure the yards of crape. Gage the bitter vintage of tears. Yes. They have more than we have. It is our fort.

We won it fairly. We are the best killers. Man to man, we can kill more of them than they can of us. That establishes the righteousness of any cause.

The night after the battle isn't so still as the night before. The soldiers are so wearied, mind and body and soul so tired, they moan a little in their sleep. A man babbles—in a strange tongue. He was the first man in the embrasure, and he is hurt in the head. He will die before morning. He is talking to his mother, who died in a little Italian mountain village when the soldier was a tiny boy—talking to her in the soft musical tongue she taught him. He hasn't spoken a word of it for many years. But he is going out of this world of misunderstandings and strife and wars, into the unmeasured years of peace. Going to God—by the way of the old home—up the winding mountain path, past the cool spring in the shadow of the great rock, through the door of the little home under the trees—such a sweet way to heaven. He is soothing the deadly pain in his head, just as he soothed all his headaches and heartaches twenty years ago, by nestling in her caressing arms and leaning his tired head against her tender breast. No; he doesn't need the chaplain. His mother is comforting him. When a man gets to his mother, it isn't very far, then, to God. A colonel sits by a camp-fire with his face in his hands. The sentinel hears him say, "O Christ!" His son was killed at his side, on the slope of the fort. The colonel has been trying to write the boy's mother. But that is harder, a thousand times harder than fighting in the death-packed embrasures. The torn sheets of paper lying like great snow-flakes about his feet are the letters he has begun. "My precious wife," "Heart of my heart," "My own heart's darling." It's a big price to pay for a dirt fort. There is a saying that "All's fair in war." But the truth is, nothing is fair in war. The winner has to pay for his winnings about as much as the loser pays for his losses. And the trouble is, neither one can pay spot cash and have the transaction over and done with. The paying for a fort goes on as long as a winner or loser is left alive—heartache and loneliness and longing and poverty and yearning and bitterness.

Takes a long, long time to pay for a common dirt fort, fairly won by fair fighting.

And then, after you've won it, and have been paying for it so many years, you haven't got it, after all.

Afterglow

Years after the battle, a journey carried me back to the field that was plowed into blood-sodden furrows by the iron shares of war's fierce husbandry. And one evening in May I walked, with my wife by my side, out of the little town to show her the fort whose name and story I had seen written in blood and fire and smoke. I had often told her that I could find the place if I were stone blind. I knew my way now. This direction from the little river—so far from the hill—this way from the stone mill. This is the sloping field, sure enough. I remember how my heart pumped itself well-nigh to bursting as I ran up the grade, shouting with the scanty breath I needed for running. And here, at the crest of the slope, was that whirlwind of flame and thunder, the Fort. Here—under our feet.

The sun was going down and all the west was ruby and amethyst set in a clasp of gold. A red-bird was singing a vesper-song that throbbed with love-notes. In the door of the cottage, garlanded with vines, a woman was lifting her happy laughing face to the lips of a man who, with his coat flung over his arm, had just come in from a field. And in merry circles, and bewildering mazes, over the velvet grasses and the perfumed violets that carpeted the sweet earth where the Fort should have stood, a group of romping children laughed and danced and ran in ever-changing plays, and all the world around the old hell-crater was so sweet and happy with peace and love and tenderness that the heart had to cry because laughter wasn't happy enough to speak its joy and gratitude. I held the hand of my dear wife close against my heart as she nestled a little nearer to my side, and I thanked God that I couldn't find the Fort I helped to win.

It was built to resist plunging solid shot and bursting shell and treacherous mine; the storm of shouting columns and the patient strategy and diligence of engineer and sapper. But God—God the all-loving Father, scattered the soft white flakes of snow—lighter than drifting down upon it, for a few winters. For a few summers he showered upon it from the drifting clouds light rain-drops no bigger than a woman's tears. He let the wandering winds blow gently over it. The sheep grazed upon its slopes. The little children romped and played over it. The clinging vines picked at it with their tiny fingers. And lo! while the soldier's memory yet held the day of its might and strength and terror, it was gone.

Benediction

"Then the same day at evening"—the evening of the first Sunday; only three days after the agony of Gethsemane; the terror of Olivet, the storm of hate and bigotry on Calvary, the blood and sacrifice, the awful tragedy of the cross, the splendor of the resurrection—"came Jesus and stood in the midst and saith unto them, 'Peace be unto you.' "

And the horror and the fear and the anguish were gone. "Then were the Disciples glad." They knew His face by the peace that shone upon it. The benediction of his lips rested on their souls. "Peace." And the storm was over. To-day, we climb the hill outside the gates of the city, and we can not find the holy spot whereon they crucified Him. We know the storm of warring human passions, of anger and bigotry and ignorance that raged around His cross. But we can not find the spot where it stood. For all the green hill is beautiful in the blessed tranquillity of the peace that endures. For love is sweeter than life, and stronger than death, and longer than hate. The hand of the conqueror and the hand of the vanquished fit into each other in the perfect clasp of friendship. The flag that waved in triumph and the flag that went down in defeat cross their silken folds in graceful emblem of restored brotherhood. The gleaming plowshare turns the brown furrow over the crumbling guns that plowed the field of life with death. God's hand has smoothed away slope and parapet of the Fort that was won for an hour and lost forever.

THE END

PRAIRIE STATE BOOKS

Mr. Dooley in Peace and in War
Finley Peter Dunne

Life in Prairie Land
Eliza W. Farnham

Carl Sandburg
Harry Golden

The Sangamon
Edgar Lee Masters

American Years
Harold Sinclair

The Jungle
Upton Sinclair

Twenty Years at Hull-House
Jane Addams

They Broke the Prairie
Earnest Elmo Calkins

The Illinois
James Gray

The Valley of Shadows: Sangamon Sketches
Francis Grierson

The Precipice
Elia W. Peattie

Across Spoon River
Edgar Lee Masters

The Rivers of Eros
Cyrus Colter

Summer on the Lakes, in 1843
Margaret Fuller

Black Hawk: An Autobiography
Edited by Donald Jackson

You Know Me Al
Ring W. Lardner

Chicago Poems
Carl Sandburg

Bloody Williamson: A Chapter in American Lawlessness
Paul M. Angle

City of Discontent
Mark Harris

Wau-Bun: The "Early Day" in the North-West
Juliette M. Kinzie

Spoon River Anthology
Edgar Lee Masters

Studs Lonigan
James T. Farrell

True Love: A Comedy of the Affections
Edith Wyatt

Windy McPherson's Son
Sherwood Anderson

So Big
Edna Ferber

The Lemon Jelly Cake
Madeline Babcock Smith

Chicago Stories
James T. Farrell
SELECTED AND EDITED BY CHARLES FANNING

The Drums of the 47th
Robert J. Burdette

University of Illinois Press
1325 South Oak Street
Champaign, IL 61820-6903
www.press.uillinois.edu